Remote Sensing of the Mine Env

Şebnem Düzgün devotes this book to her sons, Onat and Kutad, and to her family,

and

Nuray Demirel devotes this book to her husband Dursun Demirel and to her family.

The authors also devote this book to all those who dedicate their lives to science.

Remote Sensing of the Mine Environment

H. Şebnem Düzgün
Nuray Demirel

CRC Press
Taylor & Francis Group
Boca Raton London New York

CRC Press is an imprint of the
Taylor & Francis Group, an **informa** business

Cover illustrations:
Top: A typical earth observation satellite. Illustration created by Şebnem Düzgün and Nuray Demirel.
Middle: Marion 8050 Dragline, a giant machine commonly utilized for overburden stripping in surface coal mines. Photo was taken by Onur Gölbaşı on 18 October 2009 in a coal mine in Turkey.
Bottom: Landsat TM7 +ETM satellite image. Courtesy of Satellite Image Corporation.

CRC Press
Taylor & Francis Group
6000 Broken Sound Parkway NW, Suite 300
Boca Raton, FL 33487-2742

First issued in paperback 2017

CRC Press/Balkema is an imprint of the Taylor & Francis Group, an informa business

©2011 Taylor & Francis Group, London, UK

Typeset by MPS Limited, a Macmillan Company, Chennai, India

No claim to original U.S. Government works

ISBN-13: 978-0-415-87879-1 (hbk)
ISBN-13: 978-1-138-11605-4 (pbk)

Library of Congress Cataloging-in-Publication Data

Düzgün, Şebnem.
   Remote sensing of the mine environment / Şebnem Düzgün & Nuray Demirel.
      p. cm.
   Includes bibliographical references and index.
   ISBN 978-0-415-87879-1 (hardback)
   1. Mineral industries—Environmental aspects—Remote sensing.   2. Environmental impact
III. analysis.   I. Demirel, Nuray.   II. Title.

   TD195.M5D89   2011
   622—dc22

                                                      2011006896

Published by: CRC Press/Balkema
                P.O. Box 447, 2300 AK Leiden, The Netherlands
                e-mail: Pub.NL@taylorandfrancis.com
                www.crcpress.com – www.taylorandfrancis.co.uk – www.balkema.nl

Visit the Taylor & Francis Web site at
http://www.taylorandfrancis.com

and the CRC Press Web site at
http://www.crcpress.com

Contents

Preface

Remote Sensing of the Mine Environment is a guide for mining professionals, students, environmental scientists, and engineers who are interested in monitoring the environmental impacts of mining activities such as mine subsidence monitoring, slope stability monitoring, reclamation planning and implementation, and post-closure mine monitoring. Readers should have a basic understanding of environmental impacts due to mining activities and be ready to learn how to implement remote sensing technologies in monitoring them. Each chapter is attempted to be made as complete as possible, therefore, researchers, mining professionals, and students can adopt this book to their own requirements. The chapters are presented by discussing in detail the necessary principles and techniques, and briefly presenting the case studies demonstrating how to apply different remote sensing data and techniques for different mine monitoring purposes. The case studies in the book are expected to provide a comprehensive guide to applying remote sensing techniques in various aspects of the mine monitoring. The book has been planned to serve as an introductory text to the up-to-date techniques and practical applications of remote sensing on mine environment monitoring. It endeavors to fulfill the needs of students, mining professionals, and researchers on application of remote sensing for tackling problems related to environmental impacts of the mines.

The authors would like to express their thanks to various individuals and organizations for the help provided during the preparation of this book. In particular, they would like to thank to Onur Gölbaşı, Mustafa Erkayaoğlu, Mahmut Çavur for helping the authors in organizing the figures in the book. The authors are also grateful to Mustafa Kemal Emil and his supervisor H. Şebnem Düzgün; Hande Yetişen and her supervisor Mehmet Ali Hindistan; and Nuray Erdoğan Demirel and her supervisors Abdurrahim Özgenoğlu and Zuhal Akyürek for providing this book with their research works as case studies. Furthermore, sincerest thanks are extended to Germaine Seijger, who suggested the preparation of this book, for her guidance, to José van der Veer for her guidance and assistance throughout the whole process.

Finally, the authors owe special thanks to their families for their endless support and understanding throughout the preparation journey of this book.

H. Şebnem Düzgün and Nuray Demirel
Ankara, Turkey
March 2011

About the Authors

H. Şebnem DÜZGÜN is a professor in the Mining Engineering Department at the Middle East Technical University (METU). She is also affiliated to the interdisciplinary program of Geodetic and Geographic Information Technologies (GGIT) and Applied Mathematics Institute of METU. After getting a Ph.D. degree from the Mining Engineering Department at METU, Professor Düzgün did postdoctoral research in the Civil and Environmental Engineering Department at Massachusetts Institute of Technology, Cambridge, Mass., USA between 1998 and 1999 and the Norwegian Geotechnical Institute, Oslo, Norway between 2004 and 2005. She has more than 20 years of active research and teaching experience in both mining engineering and remote sensing. Professor Düzgün has developed and taught several courses on geographic information systems and remote sensing for the GGIT program, and rock mechanics, mine closure and reclamation, mine design and mine system analyses courses in the Mining Engineering Department of METU. She is the author of several book chapters on geographic information systems and remote sensing. H. Şebnem Düzgün has written more than a hundred publications on mining engineering and application of geographic information systems and remote sensing in various disciplines.

Nuray DEMİREL is an assistant professor of Mining Engineering at the Middle East Technical University (METU). She holds a Ph.D. (Missouri University of Science and Technology, formerly University of Missouri-Rolla, USA), M.Sc., B.Sc., and B.Sc. Minor Degree Diploma (Middle East Technical University, Turkey). Dr. Demirel has over ten years of professional experience in research and four years in teaching. She has taught several undergraduate and graduate courses in Mining Engineering and has graduated several students for industry and academia. Her current research areas include geographic information systems and remote sensing, surface mining, mining machine health and longevity, dragline simulation and modeling, virtual prototype, kinematics and dynamics of excavators. Dr. Demirel is leading major research initiatives in these areas. The results of her research initiatives include three books, several refereed journal and conference publications, and dissertations and technical presentations to various academic, government, and industry audiences.

## Chapter 1

Introduction

Extraction of mineral resources in a sustainable framework is vital for the welfare of human beings. The mining industry has ever been perceived negatively due to its impacts on land, water, and air at local, regional, and global levels. Minimizing the footprint of mining activities and the shift of current mining practices to a sustainable mining are key drivers towards mine environment monitoring. In this sense, remote sensing plays a significant role in mine monitoring to sustain safe and effective operations and mitigate the risks associated with them.

1.1 SUSTAINABILITY IN MINING

Increasing world population growth rate yields rising demand for finite natural resources. This results in accelerating depletion of non-renewable natural resources and makes human sustainability dubious. Public awareness and sensitivity related to environmental issues and use of scarce resources have led to sustainable development emerging as one of the major global issues since the 1980s. Sustainable development was first defined in 1987 by the Brundtland Commission of the United Nations as meeting the requirements of the present generations without compromising the ability of future generations to meet their own needs (UNGA, 1987). It was also defined as sustaining and improving living standards within a given carrying capacity of supporting eco-systems. This can only be achieved by meeting environmental, social, and economical demands concurrently (UNGA, 2005).

Sustainability concepts in mining industry are a rather more complex and controversial issue. It is argued that mining cannot truly be sustainable due to finite mineral resources. In contrast, the mining industry rejected the perception of finite mineral resources due to the discoveries of new deposits and improved technological facilities that allow mining of ore reserves, which were uneconomical in the past due to their depth, grade or other challenging factors. A growing public pressure and awareness compels the mining industry to address the issue of sustainability and to extract mineral resources within a sustainable development framework. The International Council on Mining and Metals has developed a sustainable development framework by defining a set of principles since 2001. These principles, also called The **10 principles**, required to be implemented for sustainable mining, are given by ICMM (2003):

1   Implement and maintain ethical business practices and sound systems of corporate governance.

2 Integrate sustainable development considerations within the corporate decision-making process.

3 Uphold fundamental human rights and respect cultures, customs, and values in dealing with employees and others who are affected by mining activities.

4 Implement risk management strategies based on valid data and sound science.

5 Seek continual improvement of public health and safety performance.

6 Seek continual improvement of environmental performance of mining.

7 Contribute to conservation of biodiversity and develop integrated approaches to land use planning.

8 Facilitate and encourage responsible product design, use, re-use, recycling, and disposal of products during mining.

9 Contribute to the social, economic, and institutional development of the communities around mining activities.

10 Implement effective and transparent engagement, communication and independently verified reporting arrangements with the stakeholders.

The 10 principles in fact address the three aspects of sustainability associated with mining activities as issues related to business (principles 1–3), environment (principles 4–8) and society (9–10). Essentially, sustainable development in mining industry requires consideration of declining mineral and energy resources, ore grades and recovery factors, available resources, risks, and environmental and social impacts of mining and post-mining activities (Mudd, 2007). For addressing these issues and assessing sustainability of mining industry in a factual manner, extensive data about ore grades, production rates, economic benefits, environmental burdens etc. and their systematic analysis are required. In this regard, environmental monitoring of mining activities is vital to sustainable mining practices, especially for providing safe and improved mine environments as indicated in principles 5 and 6. Systematical monitoring and assessing the impacts of mining is also crucial for investigating biodiversity conservation and developing integrated land use planning methodologies, as addressed in principle 7. Therefore, a continuous monitoring framework for the mine environment can help in sustaining the mining industry as well as the development of society. The mine environment does not only cover the area of land where mining is practiced, but also involves the surrounding environment, which is larger than the licensed mining site. As mining can have impacts to the upstream and/or downstream environment, mine environmental monitoring should cover a relatively larger land-surface than the actual area of the mine, and this is where remote sensing provides cost-effective solutions.

1.2 ENVIRONMENTAL IMPACTS OF MINING

Mining inherently causes various environmental challenges ranging from land disturbance, to water and air pollution. These challenges might potentially affect the living standards of people, habitat, and overall environment in a local, regional or global scale if not managed well (Figure 1.1). The impacts of mining activities can be grouped into three main categories according to the level of affected environmental components as (i) land, (ii) water, and (iii) air disturbances. The type and magnitude of these disturbances change depending on the mine size, amount of the production, type of mining method, and the lifespan of the mine.

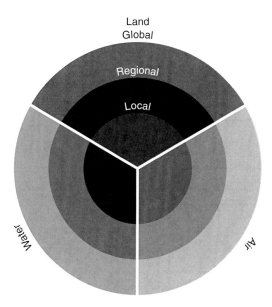

Figure 1.1 Classification of mining impacts and their extension (See color plate section)

Land disturbances

Geographical location and areal extension of mineral deposits are not controllable and mining takes place where the mineral deposits are found. Mining activities generally impose intensive land use change during exploration, development, and production stages, thus land disturbances become a principal consequence of mining. The magnitude of this disturbance depends upon the areal extent and depth of the ore deposit, and the type of mining method.

In surface mining, large volumes of overburden material are excavated and removed from one place to another causing continuous change in topography with time. Large amounts of rock and ore are stripped and stored, which consequently creates huge holes and piles in the landscape and leaves "scars" on the Earth surface (Figure 1.2). Removing top soil, which is an essential step to access and to strip overburden rock material, may result in displacement of soil, change in biodiversity, deforestation, and reduction of agricultural areas, disturbance of local water resources due to change in catchments areas and destruction of stream networks.

In addition to the land disturbance, large-scale surface mining operations can cause considerable amounts of deforestation due to the construction of the infrastructure of the mine. Another significant impact of mining activities on land is soil erosion. Although soil erosion is a natural process, mining activities can accelerate this process, the most erodible material, from one place to another when it is not removed and stored carefully. The fine sediments from the overburden dump site may be transported into the water resources after heavy rainfall.

When compared to surface mining, underground mining has less impact on land. However, mine subsidence can be a critical issue with certain underground mining

Figure 1.2 Aerial view of an open pit mine *(Source: KCGM web site, http://www.superpit.com.au/ PhotoLibrary/OpenPit/tabid/176/Default.aspx, with the permission of KCGM)* (See color plate section)

methods such as longwall mining, caving methods etc., due to the removal of ore and distraction of the natural stress state of the rock strata. As a result, the ground level lowers and surface topography changes. The land disturbance due to subsidence occurs especially in longwall coal mining as this mining method relies on excavating the coal and leaving the rock on top of the coal to cavein. In order to quantify the effects of underground mining in a particular region, a thorough understanding of subsidence patterns should be achieved.

In addition to the physical disturbance of the land in the mine site, mining activities can create chemical disturbance of the land mainly by soil pollution. The major causes of the soil pollution are:

- Acid rock/mine drainage,
- Contamination by heavy metal and leaching,
- Pollution by chemicals used for mineral processing and mine workshops.

The details of these pollution sources are explained in the water disturbance part, which also has implications for soil pollution.

Water disturbances

Considering the nature of the work and scale of operations, mining activities could also be one of the threats for water resources which are the critical and vital resource for human beings and habitat. The main impacts of mining on water resources are significant use of fresh water in mineral processing, and water pollution from discharged mine effluent and seepage from tailings and waste rock impoundments. Land disturbance also induces water disturbances. The change in surface water drainage

Figure 1.3 Acid mine drainage in an open pit copper mine (See color plate section)

networks due to land disturbance results in decreased catchment areas. In addition, land disturbance by deforestation, removal of top soil, dumping of overburden, etc. cause erosion by increasing the sediment load to the water resources. The impacts of water disturbance induced by land disturbance usually emerge over relatively long time periods. On the other hand, water disturbance due to mining activities on the water quality materializes in a relatively short period of time and can be hazardous to the habitat and environment. There are four main types of mining impacts on water quality. They are briefly described as follows:

1   *Acid Mine Drainage (AMD):* Acid mine drainage (AMD) is one of the most severe environmental impacts of mining. It can be described as a natural process in which sulphuric acid and ferrous ions are produced when sulphides in rocks, like pyrite, react with air and water and oxidize (Figure 1.3). Therefore, AMD is mostly observed in mines where a large amount of rock, containing sulphide minerals, is exploited. The acidity of the environment increases when ferrous ions are oxidized and produce a hydrated iron oxide (Sengupta, 1993). Sengupta (1993) stated that a ton of coal containing one percent pyritic sulphur has the potential of producing 15 kilograms of hydrated iron oxide and around 25–30 kilograms of sulphuric acid. The acidic water body with a low pH creates a corrosive environment and cannot support the aquatic life. Acid mine drainage occurs as long as the source rock is exposed to air and water and until the sulphides are leached out. This acid leaching process occurs by rainwater or surface drainage and acidic water being carried to the streams, rivers, lakes, and groundwater. AMD could be a persistent environmental problem that must be monitored and controlled continuously. Figure 1.3 shows an example of acid mine drainage from

Figure 1.4 Acid mine drainage from abandoned underground coal mine (See color plate section)

    a copper mine, where a small pond under the conveyor belt is formed by the collection of acidic surface water in the mine's dump site. Figure 1.4 is another example of AMD, where groundwater coming from the abandoned galleries of an underground coal mine is discharging to a ditch next to an old mine road.

2   *Heavy Metal Contamination:* Heavy metal contamination is another important environmental impact of mining for mines of heavy metal containing rocks. It is caused when heavy metals such as arsenic, cobalt, copper, cadmium, lead, silver, and zinc react with acidic water. Metals are leached out and carried to nearby streams, rivers, lakes, and groundwater as water washes over the rock surface. The existence of AMD also triggers heavy metal contamination by dissolving these metals with acidic water content.

3   *Pollutants from Mineral Processing and Mine Workshops:* Another type of pollution occurs when chemical agents, like cyanide or sulphuric acid used to gain the valuable mineral and eliminate the gangue mineral, leach from the mine site to water bodies. Improperly handled chemical agents of lubricants, and oils from the workshops of the mine site can also be transported by the surface run-off water to the water resources. The accumulation of these chemicals can create a toxic environment and threaten the wildlife (Figure 1.4).

4   *Erosion and Sedimentation:* The development stage of mining necessitates construction and maintenance of haulage roads, tailing dams, mineral processing plants, workshops, and other infrastructures. During the construction of these facilities, the natural state of soil and rock mass are disturbed. Moreover, during the mining production operations, erosion of the soil and excavated material may carry large amounts of sediments into nearby streams, rivers, and lakes, unless adequate prevention and control strategies are implemented. Excessive

Figure 1.5 Slope failures in a mine's dump site towards river beds

Figure 1.6 Erosion of partially vegetated dump site of an abandoned coal mine

sediment can clog riverbeds and change watersheds, vegetation, wildlife habitats, and aquatic organisms. Figure 1.5 is an example of failure in a mine's dump site towards the bottom of the valley with river. The failed waste material in Figure 1.5 carries a lot of sediments to the river. Figure 1.6 shows the partially forested dump site of an abandoned coal mine, where the sediments from the waste dumps are continuously carried to the river network by the surface waters.

Air disturbances

Mining activities adversely affect air quality in various ways. The most commonly observed impacts are dust, air blast, chemical pollutants, fugitive particulate matter, and gaseous emissions. Air dust can be caused by various sources regardless of the mining method, such as off-highway trucks on haul roads, unvegetated spoil surfaces, topsoil stockpiles, and ore stockpiles. Air dust is especially a problem in areas of low rainfall, high winds, erodible soils, and fine waste rock. In addition to this, air impacts such as noise and air shock, caused by overburden blasting, can be dangerous if people live within a radius of several kilometers of the blasting site. The magnitude of the problem depends on several factors, such as, the type and depth of overburden being blasted, amount of explosives, the powder factor, and the distance from the mine site.

The impacts of mining on the environment and the required level and scale of monitoring for developing mitigation measures for these impacts are given in Table 1.1. As can be seen from Table 1.1, long temporal scales of monitoring are required for land and water disturbances as compared to air disturbances. Similarly, larger areas need to be monitored for land and water disturbances (Table 1.1).

1.3 ENVIRONMENTAL IMPACT MONITORING IN MINING

An increased focus on global warming and sustainable development necessitates mining engineers and decision makers having to prioritize environmental responsibility while maximizing economic benefits to the industry. In an effort to accomplish this, mine monitoring is essential to evaluate environmental aspects and potential impacts associated with mining activities and their upstream and downstream processes. Environmental impacts and associated risks can only be mitigated by implementing some safeguards and monitoring the mine environment continuously.

Monitoring programs can be established for a number of purposes (Spitz and Trudinger, 2009):

- To detect and analyze differences in physical, chemical, and biological attributes in the mine environment and its surroundings like quality of water bodies and soil during various mining activities.
- To confirm the achievement of planning objectives which are set at the beginning of the mine planning stage like measuring the sediment load to the water bodies to justify the goal being set at no or very little sediment transportation.
- To determine impacts posing risk to the environment and to identify remedial measures to decrease the risk, like monitoring the stability of tailing dams and assessing the related safety levels.
- To ensure that emissions and discharges comply with the established standards related to environmental impacts of mining activities like keeping the pH value of the discharged water from the tailing dam according to EPA standards.
- To identify the efficiency of mining activities and set the level of achievements to the goals like assessing the total amount of water usage, discharge and recycling.
- To understand interaction between various environmental impacts such as level of metal contamination in the soil and in crops, the metal content of the water in a river and that in flora and fauna of the river.

Table 1.1 List of impacts and essential level of monitoring for different types of mining methods

	Land		Water		Air	
	Impact	Monitoring Scale/Level	Impact	Monitoring Scale/Level	Impact	Monitoring Scale/Level
Open Pit and Strip Mining	• Land use change • Removal of top soil • Removal of subsoil • Huge holes and scars on the Earth's surface • Deforestation • Reduced agricultural area	• Regional monitoring on the basis of years and decades	• Acid mine drainage • Heavy metal contamination of water resources • Extensive use of water • Chemical contamination • Drainage network destruction and transportation of sediments	• Regional monitoring on the basis of months to years	• Dust • Airblast • Particulate matter • Emissions to atmosphere	• Local monitoring on the basis of days to months
Quarrying Mining	• Land use change • Removal of top soil • Removal of subsoil • Deforestation • Reduced agricultural area	• Local monitoring on the basis of years	• Extensive use of water • Chemical contamination • Drainage network destruction and transportation of sediments	• Regional monitoring on the basis of months to years	• Dust • Airblast	• Local monitoring on the basis of days to months
Underground Mining	• Mining subsidence • Alteration of natural stress state of rock mass	• Local monitoring on the basis of months to years to decades	• Acid mine drainage • Heavy metal contamination of groundwater	• Regional monitoring on the basis of years	• Dust to surface stockpiles	• Local monitoring on the basis of days to months basis

- To assess if the environmental impacts are due to mining activities or other manmade/natural activities, such as if the sediment load to water bodies is due to erosion from the mine environment or other activities around the water bodies.
- To investigate the performance of remedial measures, like assessing the effect of lime implementation to acidic waters and forming a relation between the amount and content of lime with the pH level of the water.
- To collect data related to claimed liabilities like measuring seismicity after blasting around buildings near mine sites to investigating the cracks in buildings, which are claimed to have formed due to blasting.
- To evaluate the performance of rehabilitation activities like monitoring revegetation health during progressive reclamation of a mine site.
- To establish a baseline for auditing like collecting baseline data prior to mine operations.

1.4 MINE LAND MONITORING

Mine land monitoring is of paramount concern to mine owners, mine planners, environmental scientists and engineers, policy makers, and all other decision makers who are interested in investigating the changes in mine environments. Mine land monitoring is required at each stage of mining activities for different purposes. Prior to mining activities, a landuse map of the mine area and its surroundings is usually generated for the baseline study. During mining activities, mine land monitoring might be utilized for different purposes, such as, mineral exploration, performance assessment of progressive reclamation, stability of the pit and overburden dump sites, level of induced subsidence etc. Once the mining is finished, implementation of a mine closure plan, rehabilitation practices, and their control again necessitates mine land monitoring.

There have been various techniques used in mine land monitoring for different purposes such as surveying and mapping, field measurements, and remote sensing methods coupled with field data collection. This book is focused on remote sensing techniques utilized in mine environment monitoring and the next section will briefly define the role of remote sensing in mine monitoring.

1.5 ROLE OF REMOTE SENSING IN MINE MONITORING

Mine sites are located where the mineral resources exist and they often have difficult access. Instrumentation, sampling, surveying, and photogrammetric measurements in the field necessitate teams of experts working for long periods of time. Therefore, data collection and visual inspection through field studies may become very expensive, labor intensive, and time consuming. On the other hand, remote sensing provides data for a large area in a shorter time and an increasing number of sensors and satellites results in economically affordable up-to-date information about the processes taking place on the land. Therefore, remote areas with difficult access can be monitored using land cover maps derived by remote sensing. Also, an aptitude to cover a large surface area in a very short time is another important merit of remote sensing. Conventional techniques like surveying and topographic measurements may not be practical for large areas.

Remote sensing also allows one to detect features which are not visible by human eye, and important in mine site monitoring for discovering large scale geological structures, i.e. faults, alteration zones, vegetation health, pollution etc. It has advantages in combining the findings with other geographic features integrated with geographic information systems. Nowadays there are numerous optical, radar, and laser sensors providing various spatial, spectral, radiometric, and temporal resolutions that can effectively be utilized for mine monitoring purposes. Increasing numbers of satellite-based, airborne, and terrain-based sensors makes remotely sensed data relatively cheap and also decreases the dependency on a few data sources.

Moreover, in case of lack of past data and a baseline study, archived remotely sensed data provides an opportunity to generate a landuse map and baseline data for the past. Other key benefits of remote sensing in mine monitoring can be enabling regular visits to area under study by programming the satellites and serving a large archive of historical data with continuous data acquisition.

1.6   SCOPE OF THE BOOK

This book is composed of six successive chapters to provide readers with fluent and comprehensive explanations to cover mine monitoring techniques using remote sensing with complementary case studies. Following this introductory chapter, Chapter 2 presents the principles of remote sensing. In Chapter 2, basic definitions, principles of electromagnetic radiation, the main physical laws related to electromagnetic radiation, characteristics of the electrometric spectrum, and interaction of electromagnetic energy with the atmosphere and the surface features of the Earth are presented. Following Chapter 2, Chapter 3 provides readers who are not familiar with remote sensing, with the nature and properties of remotely sensed data as well as its processing and interpretation techniques to get them ready to follow the rest of the book completely. Chapter 4 presents the details of mine subsidence and its monitoring using remotely sensed satellite imagery, aerial photographs, and radar data. It also provides a case study performed to monitor the underground coal mine subsidence using aerial photographs including processing of aerial photographs. Chapter 5 describes comprehensively mine slope stability monitoring using remote sensing data. Problems with slope stability, instrumentation, measurement, and monitoring of slope instabilities, methods of remote sensing in slope stability monitoring and its processing are explained in detail. A case study, the monitoring of slopes in an abandoned mine area using remote sensing data, is presented. Finally, Chapter 6 provides techniques for mine reclamation and post-closure mine monitoring with case studies.

REFERENCES

ICMM, International Council on Mining and Metals. (2003) http://www.icmm.com/our-work/sustainable-development-framework/10-principles.

Mudd, G.M. (2007) An assessment of the sustainability of the mining industry in Australia. *Australian Journal of Multi-Disciplinary Engineering*, 5 (1), 1–12.

Sengupta, M. (1993) Environmental impacts of mining, monitoring, restoration, and control. USA, Lewis Publishers.

Spitz, K. & Trudinger, J. (2009) *Mining and the Environment from Ore to Metal*. London, Taylor and Francis.

United Nations General Assembly. (1987) *Report of the World Commission on Environment and Development*, General Assembly Resolution 42/187, 11 December 1987. (Retrieved 14 November 2007).

United Nations General Assembly. (2005) *2005 World Summit Outcome*, Resolution A/60/1, adopted by the General Assembly on 15 September 2005. (Retrieved 17 February 2009).

Chapter 2

Principles of remote sensing

This chapter covers fundamentals of remote sensing. It provides the basic definitions, principles of electromagnetic radiation, the main physical laws related to electromagnetic radiation, and characteristics of the electrometric spectrum. The chapter then gives interactions between electromagnetic energy and the atmosphere and surface of the Earth features which will be needed to follow the remaining parts of this book.

2.1 DEFINITION OF REMOTE SENSING

Remote sensing can be defined as studying the Earth's surface by collecting data and analyzing it for extracting information about a phenomenon or object without physical contact. Various comprehensive definitions of remote sensing are listed in Table 2.1.

Remote sensing in the environmental monitoring context can be described as acquiring data from areas on the Earth either in the form of images or signals obtained by measuring and recording the electromagnetic radiation from the area. It is one of the powerful means of acquiring up-to-date information on changes, activities, and processes on the Earth. Remote sensing becomes advantageous for obtaining information for large areas, where data collection is costly. It is also one of the most convenient ways of data collection for sites which are inaccessible or have difficult access.

Remote sensing builds upon principles of electromagnetic radiation. Although there are various types and implementations of remote sensing, which will be discussed in the further sections, the way of acquiring data through remote sensing can be best understood by grasping the interaction of electromagnetic radiation with matter through radiation laws.

2.2 PRINCIPLES OF ELECTROMAGNETIC RADIATION

Radiation can be defined as energy coming from a source in the form of waves. This energy cannot be emitted by the source at absolute zero, which is $-273°C$ or $-459°F$. In fact all types of objects have a certain amount of radiation and the objects also reflect radiation coming from other objects. Remote sensing builds upon measuring such reflected and emitted energy from the objects. Obtaining information from remotely sensed data relies on an understanding of the interaction between the electromagnetic radiation and the objects or areas on the Earth.

Table 2.1 Selected remote sensing definitions

Reference	Definition
Holz (1973)	"… sensing an object or phenomenon without having the sensor in direct contact with the object being sensed."
Fussel et al. (1986)	"Remote sensing is the noncontact recording of information from the ultraviolet, visible, infrared, and microwave regions of the electromagnetic spectrum by means of instruments such as scanners and cameras located on mobile platforms, such as aircraft or spacecraft, and the analysis of acquired information by means of photo interpretive techniques, image interpretation, and state-of-the-art image processing systems."
Schowengerdt (1997)	"… the measurement of object properties on the earth's surface using data acquired from aircraft and satellites."
Milman (1999)	"… the science of measuring the properties of objects we cannot touch, by measuring the radiation they emit or scatter, or the acoustic waves they transmit."
Lillesand and Kiefer (2000)	"*Remote sensing* is the science and art of obtaining information about an object, area, phenomenon through the analysis of data acquired by a device that is not in contact with the object, area, or phenomenon under investigation."
Aranof (2005)	"… the science, technology and art of obtaining information about objects from a distance …"
Cracknell and Hayes (2007)	"Remote sensing may be taken to mean the observation of or gathering of information about a target by a device separated from it by some distance."
Campbell (2008)	"Remote sensing is the practice of deriving information about the Earth's land and water surfaces using images acquired from an overhead perspective, by employing electromagnetic radiation in one or more regions of the electromagnetic spectrum, reflected or emitted from the Earth's surface."

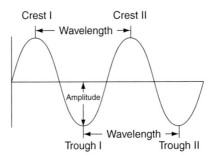

Figure 2.1 Wavelength and amplitude of a typical wave

The transfer of energy of an object to another object is governed by electromagnetic waves, which have three characteristics: Wavelength, amplitude, and frequency. *Wavelength* is defined as the distance between successive crests or trough (Figure 2.1). The unit for wavelength is expressed in meters or other order of magnitudes of meter such as *nanometers* (nm, 10^{-9} meters), *micrometers* (μm, 10^{-6} meters) or centimeters

Table 2.2 Measurement units for electromagnetic wave characteristics

Wave Characteristic	Expression	Unit
Wavelength	Distance	Kilometer (km) = 1000 m
		Meter (m) = 1 m
		Centimeter (cm) = 10^{-2} m
		Millimeter (mm) = 10^{-3} m
		Micrometer (μm) = 10^{-6} m
		Nanometer (nm) = 10^{-9} m
		Ångstrom (Å) = 10^{-10} m
Frequency	Cycles/second	Hertz (Hz)
		Kilohertz (kHz) = 10^{3} Hz
		Megahertz (MHz) = 10^{6} Hz
		Gigahertz (GHz) = 10^{9} Hz

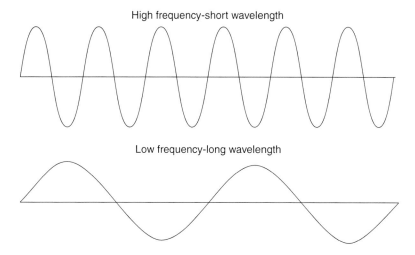

High frequency-short wavelength

Low frequency-long wavelength

Figure 2.2 Frequency of waves

(cm, 10^{-2} meters). Other variations of wavelength units are listed in Table 2.2. *Amplitude* is the height of crest or trough and it is a measure for energy of wave per its wavelength. For this reason the unit of amplitude is usually expressed as watts/m^2/μm. **Frequency** is the number of crests or troughs that pass a fixed point for a certain time period. The unit of frequency is hertz, in which 1 hertz is equivalent to 1 cycle/second (Figure 2.2).

The relation between wavelength (λ) and frequency (v) is given by Eq. 2.1.

$$c = \lambda v \tag{2.1}$$

In Eq. 2.1, c is the speed of light which is 3×10^8 m/sec. As can be seen from Eq. 2.1 electromagnetic energy has constant speed equals to speed of light and can be

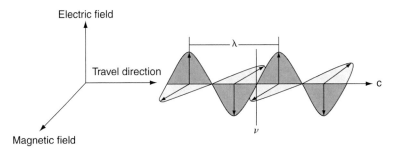

Figure 2.3 The components of electromagnetic energy

expressed either using the wavelength or frequency. The choice of expressing electromagnetic energy changes depending on the discipline. In remote sensing practice, use of wavelength in micrometers is mostly common. The wavelength and the frequency are inversely proportional, meaning that the shorter the wavelength, the higher the frequency or vice versa (Figure 2.2). The relation given in Eq. 2.1 comes from the principles of electromagnetic theory, developed by *James Clerk Maxwell* who lived between 1831 and 1879 and explained how the energy from one object to another object is transferred. The successive change in electric and magnetic fields cause both electric and magnetic waves to move and the speed of movement is equal to the speed of light. Therefore, for an electromagnetic radiation, two types of fields namely, electric and magnetic fields should exist at the same time to trigger themselves to move the waves. The components of electromagnetic radiation are illustrated in Figure 2.3. In this way electromagnetic energy from one object to another is transferred through waves of electric and magnetic fields traveling with a constant speed, which is the speed of light.

2.3 PHYSICAL LAWS OF ELECTROMAGNETIC ENERGY

The interaction of electromagnetic energy with the objects is governed by certain physical laws. *Max Planck*, who lived between 1858 and 1947, explained that the emission and absorption of electromagnetic energy by objects is not continuous but discrete. Later such a discrete unit of energy is called **quanta**. Planck proposed that the amount of radiated energy (E) is proportional to the frequency (v) of the energy and introduced a constant (h) to define proportionality of v and E (Eq. 2.2).

$$E = hv \tag{2.2}$$

The Planck's constant equals to $6.62606896(33) \times 10^{-34}$ Joule-second (J s). When frequency (v) in Eq. 2.2 is replaced by c/λ from Eq. 2.1, the Planck equation can also be expressed as in Eq. 2.3

$$E = \frac{hc}{\lambda} \tag{2.3}$$

The Eqs. 2.1–2.3 explain the transfer of electromagnetic energy between objects and the amount of energy emitted and absorbed. It is also known that electromagnetic energy emitted and absorbed by the objects is related to the temperature of the object. The association of electromagnetic energy with temperature is accounted by the notion of **blackbody** which is a theoretical and idealized object absorbing all the incoming electromagnetic energy. In real life there is no object or material which absorbs all the energy. For this reason, in real life all objects are called **graybodies**. One of the closest materials to the blackbody is graphite (an allotrope of carbon), which absorbs approximately 97% of incoming energy. *Kirchhoff's law* is based on the notion of blackbody and states that for all blackbodies at a given temperature, relative emission potential of an object's surface, called emissivity (ε), of an object is equal to its ratio of absorbed radiation for a certain wavelength called absorptivity. Hence the emissivity of a graybody is the ratio of the emittance of the graybody (M) to emittance of blackbody (M_b) at the given temperature (Eq. 2.4)

$$\varepsilon = \frac{M}{M_b} \tag{2.4}$$

The value of ε ranges between 0 and 1; 1 being the emissivity of the perfect blackbody and 0 being that of a whitebody which is 100% reflecting the electromagnetic energy.

The relation between temperature (T) and total emitted radiation (W), which is expressed in Watts/cm^2, is established by the *Stefan-Boltzmann Law* given in Eq. 2.5.

$$W = \sigma T^4 \tag{2.5}$$

In Eq. 2.5, σ is the Stefan-Boltzman constant which is equal to 5.6697×10^{-8} Watts/m$^2 \cdot$K^{-4}. According to the Stefan-Boltzman law, the fourth power of the blackbody's temperature is proportional to the total energy radiation coming from the unit area of the blackbody in unit time. The Stefan-Boltzman law implies that the radiation emitted/unit area from the hotter blackbodies is higher than that from the cooler ones.

Finally, the last law to be mentioned related to interaction between the objects and electromagnetic radiation is *Wien's displacement law* which defines the association between temperature of the blackbody and the emitted radiation's wavelength (Eq. 2.6).

$$\lambda = \frac{2897.8}{T} \tag{2.6}$$

In Eq. 2.6, T is the absolute temperature in Kelvin (K). Wien's displacement law implicates that the conversion of wavelength from longer to shorter ones, makes the blackbody's temperature increase. In other words, increasing surface temperature changes the electromagnetic energy of the blackbody, which results in color change.

The laws which are associated with the interaction of electromagnetic energy and the objects are summarized in Table 2.3.

Table 2.3 Summary of physical laws for explaining the behavior of electromagnetic energy

Name of Law	Expression	Defined Association
Maxwell	$c = \lambda \nu$	Wavelength (λ), frequency (ν), speed (c) of electromagnetic energy
Planck	$E = h\nu$	Proportionality of electromagnetic energy (E) to frequency (ν) with Planck's constant (h)
Kirchhoff	$\varepsilon = \frac{M}{M_b}$	Electromagnetic energy emittance of the graybody (M) and emittance of blackbody (M_b)
Stefan Boltzman	$W = \sigma T^4$	Proportionality of temperature (T) and total emitted radiation (W) with Stefan-Bolzman constant (σ)
Wien's Displacement	$\lambda = \frac{2897.8}{T}$	Temperature of the blackbody (T), the emitted radiation's wavelength (λ)

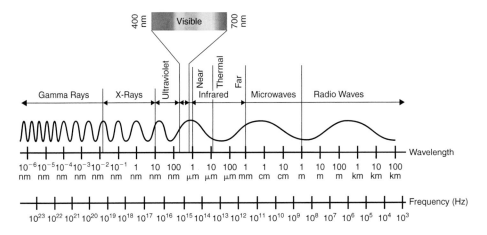

Figure 2.4 The electromagnetic spectrum (See color plate section)

2.4 THE ELECTROMAGNETIC SPECTRUM

The series of electromagnetic wavelength range forms the **electromagnetic spectrum.** Although ranges of wavelength define certain characteristics, the ranges are not divided up in a precise manner. In fact the electromagnetic spectrum is expressed in the form of continuous wavelength regions. The spectrum starts from long wavelengths, low energy radio waves and extends to short wavelengths, high energy gamma rays. Typical wavelength ranges are illustrated in Figure 2.4. The microwave, infrared, and visible portion of the electromagnetic spectrum is mostly used by the remote sensing sensors in Earth observation. A more detailed division of these portions are given in Table 2.4.

As there is an inverse relation between the energy of the quanta and its wavelength (see Eq. 2.3) by the implication of Planck's electromagnetic theory, the higher energy levels are obtained in shorter wavelengths. It is relatively more difficult to sense microwaves emitted from an object on the Earth due to low energy levels. On the other hand, it is easier to detect visible and infrared portion of the electromagnetic spectrum because of the associated higher energy levels.

Table 2.4 Ranges of electromagnetic spectrum used for remote sensing of the Earth

Region		Wavelength Range
Visible	Blue	0.4–0.5 μm
	Green	0.5–0.6 μm
	Red	0.6–0.7 μm
Infrared (IR)	Near IR	0.7–1.3 μm
	Mid IR	1.3–3 μm
	Thermal IR	3–14 μm
Microwave		1 mm–1 m

Remote sensing relies on detecting the electromagnetic energy emitted/reflected from the object, which requires interaction of the objects with energy sources. In the remote sensing context, the Sun is one of the governing electromagnetic energy sources and can be considered as a blackbody. Based on this analogy, the wavelength of the emitted electromagnetic energy from the objects on the Earth (e.g. rock, soil, forest, ocean, building, etc.) can be sensed through estimating the temperature of the object's surface by using Wien's displacement law (Eq. 2.5). On the average it can be assumed that the Earth's surface temperature, also called ambient temperature, is around 27°C (300 K). Then the corresponding wavelength for emitted energy from the Earth's surface is approximately 9.66 μm (2897.8/300 from Eq. 2.5) which is in the thermal infrared (IR) range of the electromagnetic spectrum (Table 2.4). The wavelength of thermal IR range can be detected by radiometers. As the name implies, the visible range in the electromagnetic spectrum is detectable by the human eye or camera, where the reflected solar energy within the wavelength range of 0.4–0.7 μm (Table 2.4) from the objects on the Earth is sensed. When the range of wavelength for emitted electromagnetic energy (e.g. IR range in Table 2.4) is compared with that for reflected electromagnetic energy (e.g. visible range in Table 2.4), it is clear that emitted energy has longer wavelength with higher energy level. For this reason, information related to thermal features of the objects (moisture content, coal fires, etc.) on the Earth is obtained by sensing IR wavelengths.

In addition to solar energy as an electromagnetic energy source, radar systems with microwave wavelength range (Table 2.4) are also used for remote sensing. In radar systems the microwave wavelengths are generated by the system itself and then the reflected and emitted energy from the objects are recorded. As the system creates its own energy, these systems are called **active remote sensing systems. Passive remote sensing systems,** on the other hand, are based on the solar energy, since the system cannot use its own energy but it is dependent on the existence of solar energy.

2.5 THE INTERACTION OF ELECTROMAGNETIC ENERGY WITH THE ATMOSPHERE

Whatever the electromagnetic source (radar or solar), the information collection related to the Earth's surface through remote sensing requires understanding of the interaction of electromagnetic energy with the atmosphere as the energy travels within the atmosphere. If the remote sensing sensor is mounted on an aircraft, the path and length of electromagnetic energy traveling through the atmosphere will be relatively negligible. This results in considerably better data quality. However, if the remote sensing sensor is mounted on a satellite-based platform, the electromagnetic energy will pass through the whole atmosphere. In this case, the obtained data contain atmospheric effects which should be removed before obtaining the required information. When electromagnetic energy is subjected to travel through the entire atmosphere, it changes its characteristics due to three factors:

- Scattering,
- Absorption, and
- Transmission.

The change in the direction of the energy is called **scattering**. The directional change occurs due to the suspended particles (e.g. dust, smoke, water vapor etc.) in the atmosphere. There are several factors that affect the degree of scattering. The size and density of the suspended particles, the wavelength of the electromagnetic energy and the thickness of the atmosphere in which the energy travels are the most influencing factors. The interaction of the electromagnetic radiation with the suspended particles in the atmosphere causes three different types of scattering namely: Rayleigh scattering, Mie scattering, and Nonselective scattering. The Rayleigh scattering occurs when the size (diameter) of suspended particles in the atmosphere is considerably smaller than the wavelength of the radiation. There is an inverse proportionality between the amount of Rayleigh scattering and the fourth power of the wavelength. For this reason, the short wavelengths have higher tendency to be scattered than the long ones. For example the wavelength blue is scattered four times higher than the red, ultraviolet is scattered four times higher than the blue and 16 times higher than the red. As the wavelength of blue from the solar energy traveling through the atmosphere is scattered more than the other visible wavelengths of solar energy due to the Rayleigh scattering effect, the sky has a blue color (Figure 2.5). When the distance traveled by the electromagnetic energy in the atmosphere is increased, the degree of Rayleigh scattering decreases. Hence, during sunset and sunrise, in which the solar radiation travels longer through the atmosphere, the longer wavelengths of red and orange are scattered more. As there is no interaction between the solar energy and the atmosphere in space, pictures of the Earth taken from space do not have blue sky but black sky. Rayleigh scattering is mostly dominant in the atmosphere up to the altitudes of 9–10 km and occurs when the atmosphere is free from impurities such as water vapor, dust, etc. When the atmosphere contains large size suspended particles such as water vapor, dust, pollen, smoke, etc., which have approximately the same size as the wavelengths of the visible and near infrared spectrum, Mie scattering occurs. Although Mie scattering is also dependent on the wavelength, there are other factors such as particle size, shape, density, etc. affecting the degree of Mie scattering. As the lower altitudes of atmosphere (0–5 km) involve such particles, Mie scattering occurs in lower atmosphere. Nonselective scattering occurs when the size (diameter) of suspended particles are considerably larger than the wavelength of the radiation. Water vapor in the atmosphere with a diameter ranging between 5 and 100 μm causes equal scatterings of radiation irrespective of the wavelength.

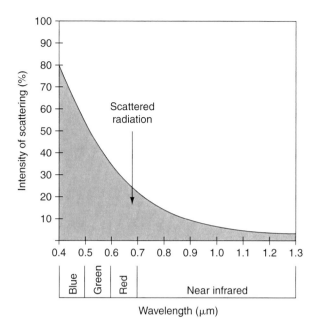

Figure 2.5 Relation between Rayleigh scattering and wavelength

As all visible and near infrared wavelengths are scattered, the clouds and fog have white color.

Scattering influences the performance of remote sensing in three different ways. First of all, due to Rayleigh scattering, the blue and ultraviolet region in the electromagnetic spectrum is considered to be useless for remote sensing, as these wavelengths detected by the remote sensing sensor are brightness of the sky rather than the brightness of the objects on the Earth. Due to this season, some of the remote sensing sensors do not have sensors for short wavelengths for blue and ultraviolet. Secondly, as scattering changes the direction of radiation, the remote sensing sensor can detect radiation from objects which are not in the viewing zone of the sensor, so-called **instantaneous field of view** (IFOV) of the sensor. Therefore, the energy recorded by the remote sensing sensor may not belong to the radiation from what the sensor is viewing, and thus, the spatial detail recorded by the sensor decreases or becomes misleading. Moreover, scattering diminishes the contrast by causing dark objects to be detected brighter than their original form or bright objects to be detected darker than their original form.

The loss of energy due to certain atmospheric gases is called **absorption**. There are three major gasses causing absorption, namely ozone (O_3), carbon dioxide (CO_2), and water vapor (H_2O). Ozone is densely concentrated at the altitude of 20–30 km in the atmospheric layer called the stratosphere as it forms due to the reaction of oxygen (O_2) and ultraviolet energy coming from solar radiation. In fact, ozone absorbs short wavelengths (less than 0.24 μm) of the electromagnetic spectrum, which are mainly the ultraviolet portion of the spectrum. By this way, the transmission of short wavelengths to the lower level of atmosphere is blocked providing suitable conditions for

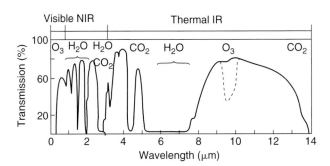

Figure 2.6 Atmospheric windows (modified from http://www.fas.org/irp/imint/docs/rst/Sect9/Sect9_1 .html)

living organisms on the Earth. Carbon dioxide exists in the lower atmosphere and absorbs mid and far infrared wavelengths. The wavelengths ranging from 13 to 18 μm are the highest absorbed wavelengths by carbon dioxide. Water vapor is also present in the lower atmosphere and absorbs more than 80% of energy in the wavelengths between 5.5–7.0 μm and wavelengths longer than 27 μm. The concentrations of ozone (0.07 ppm in the air and 0.1–0.2 ppm in the stratosphere) and carbon dioxide (0.039% of the air) are relatively constant in the atmosphere. However, the concentration of water vapor is variable depending on the time and location. The locations in humid climates, e.g. rain forests, have the highest water vapor concentration, whereas locations in the arid lands, e.g. deserts, have the lowest concentrations. As the electromagnetic energy is not directly transferred to the Earth's surface and some of the wavelengths are blocked by the three gasses, remote sensing, especially passive remote sensing based on solar radiation, uses the wavelengths of the electromagnetic spectrum transferred without absorption. The wavelengths which are not subjected to absorption are called **atmospheric windows** (Figure 2.6).

The electromagnetic energy directly passing through atmospheric windows is called **transmission**. The 3% of total electromagnetic radiation with wavelength less than 4.5 μm coming from the Sun is absorbed by stratosphere due to reaction of oxygen with ultraviolet wavelengths to form ozone. Approximately 26% of this radiation is reflected by the clouds, 20% of it is absorbed by the atmospheric gasses, 43% of it is absorbed by the Earth's surface and the remaining 8% is reflected from the objects on the Earth's surface (Figure 2.7). The ability of an object with a certain thickness to transmit electromagnetic energy is expressed by its **transmittance**. Transmittance is defined by the ratio of transmitted energy to the incident (incoming) energy. The transmittance for objects depends on the thickness of the object and the wavelength of the incident energy.

The energy absorbed by the objects on the Earth's surface is emitted again. For this reason, if 100 units of energy come to the Earth's surface 113 units are emitted to space. Recall that the emitted energy from the Earth's surface has wavelength of 9.66 μm (2897.8/300 from Eq. 2.5) according the Wien's displacement law, which is in the thermal infrared portion of the electromagnetic spectrum (Table 2.4). As the

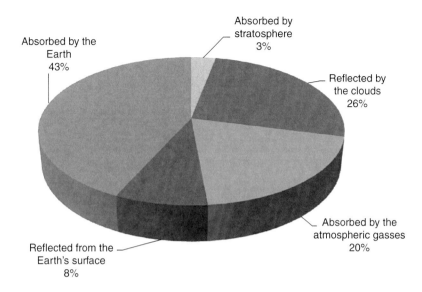

Figure 2.7 The distribution of total incoming solar energy with wavelength less than 4.5 μm

atmospheric gasses of carbon dioxide and water vapor absorb longer wavelengths, majority of the energy emitted by the Earth itself is trapped in the atmosphere. The other remaining parts results in two main processes. The first one causes heating of the lower atmospheric levels by the Earth's surface causing replacement of hot air with cooler air. The second process is evaporation of objects on the Earth's surface, e.g. evaporation of water, and from soil and vegetation cover. The part of electromagnetic energy directly transmitted in the atmosphere through the atmospheric windows is used for passive remote sensing sensors.

2.6 THE INTERACTION OF ELECTROMAGNETIC ENERGY WITH OBJECT SURFACES ON THE EARTH

Once the electromagnetic energy passes through the atmosphere, it interacts with the surfaces of objects on the Earth. This interaction results in three processes, namely reflection, absorption, and/or transmission. The incident energy, which is the energy interacting with the object's surface, is sum of the reflected, absorbed, and transmitted energy due to the energy conservation law. The degree of occurrence of these three processes is dependent on the surface characteristics of the object, wavelength of the energy incident and the angle at which the object surface is illuminated by the source.

Absorption and transmission of the electromagnetic energy has already been described in Section 2.5. Note that absorption of electromagnetic energy by the surface of the object raises the temperature of the surface and it is reemitted as heat energy by the object.

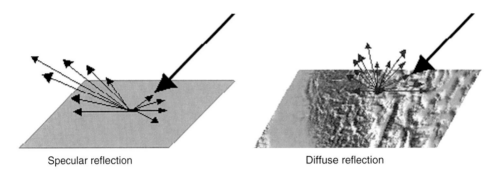

Specular reflection   Diffuse reflection

Figure 2.8 Specular and diffuse reflections

Reflection is defined by the redirection of light after hitting an opaque surface. The amount of reflection depends on the surface roughness and the wavelength of the incident energy. The surface roughness (asperity heights of the surface roughness), which is less than the wavelength, results in specular reflection. Since the surface is smoother than the wavelength in specular reflection, almost all the incident energy is reflected to a single direction (Figure 2.8). In specular reflection, the incident angle of the electromagnetic energy (the angle at which the energy hits the surface) is equal to the angle of reflection. Specular reflection occurs from surfaces, which are almost perfectly smooth, like mirrors, polished metals, and smooth water bodies. The reflection occurring from objects having surface roughness greater than the wavelength of the incident energy is called diffuse reflection. A perfect diffuse reflection causes all the energy to be scattered equally in all directions (Figure 2.8). Object surfaces having diffuse reflectance properties are also called Lambertian Surfaces as the work of Johann Lambert who lived between 1728–1777 forms the basis of diffuse reflection.

The majority of the objects on the Earth have both diffuse and specular reflection properties. Diffuse reflection determines the spectral characteristics of the object surfaces. In the visible range of the electromagnetic spectrum such spectral characteristics are called **color**. The objects are red when they reflect the red portion of the visible range in the electromagnetic spectrum (0.6–0.7 μm in Table 2.4). Diffuse reflection also determines the way the objects are viewed in different wavelengths. A beach with fine grain sand usually seen as a smooth surface in microwave wavelengths (1 mm–1 m in Table 2.4), under active remote sensing conditions, whereas the same beach will have quite rough surface in visible portions under passive remote sensing conditions. For this reason diffuse reflection has particular importance in remote sensing, which is mainly based on detecting the diffuse reflection characteristics of objects on the Earth. The reflectance characteristics of object surfaces [$\rho(\lambda)$] for a given wavelength (λ) are measured by the ratio of reflected energy [$E_r(\lambda)$] from the object surface to the incident energy on the object surface [$E_i(\lambda)$] (Eq. 2.7).

$$\rho(\lambda) = \frac{E_r(\lambda)}{E_i(\lambda)} \tag{2.7}$$

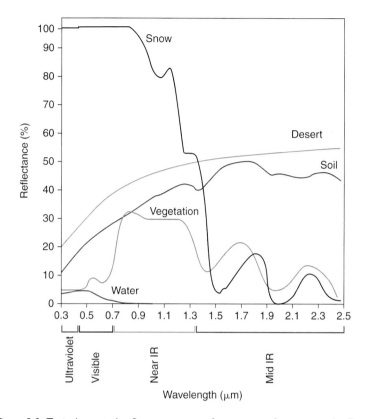

Figure 2.9 Typical spectral reflectance curves for common features on the Earth

$\rho(\lambda)$ also defines the **spectral reflectance** of an object surface. When $\rho(\lambda)$ is obtained for various wavelengths, spectral reflectance curves are obtained. These curves constitute the reflectance signature of Earth features so that they provide appropriate wavelength values to be used in remote sensing applications. Figure 2.9 shows spectral reflectance curves for water, desert, soil, snow, and vegetation. As can be seen from Figure 2.9, each feature has its own characteristics at different wavelengths. For example, while snow has the highest reflectance for the wavelength range of 0.3–1 µm and the lowest reflectance for the wavelength of 1.5 and 2.0 µm, water has no reflectance for the wavelengths greater than 0.8 µm. Note that such reflectance characteristics are for idealized features. If snow and water have impurities, for instance, depending on the type of impurities the reflectance characteristics will have different curves.

In general, the amount of electromagnetic energy absorbed, transmitted, and reflected by the objects on the Earth, changes according to the object type and its surface properties. Remote sensing benefits from this variable nature of electromagnetic energy interaction with the surfaces for distinguishing different features. Remote sensing also utilizes the wavelength dependency of absorption, transmission, and reflection properties of the Earth features. By this way two or more inseparable features can be distinguished by viewing them in different wavelengths. Recall the example of beach with fine grain sand detected by active and passive remote sensing systems.

Remote sensing sensors detect and record reflected, absorbed or emitted electromagnetic energy from the objects which are in the line-of sight or IFOV of the sensor. Most of the remote sensing sensors are based on electro-optical principles and thus record reflected energy, mostly various ranges of the electromagnetic spectrum. The emitted heat from the objects due to interaction of electromagnetic energy with the object surfaces are collected by radiometers that record mostly the infrared portion of the electromagnetic spectrum. The sensors used in meteorological studies, on the other hand, record absorption of electromagnetic energy related to carbon dioxide (CO_2), ozone (O_3), and water vapor (H_2O). Moreover, as the microwave portion of the electromagnetic spectrum allows direct transmission of the energy, the active remote sensing sensors like radars provide detection of clouds and precipitation, as well as detecting Earth features such as surface roughness, moisture content, etc.

2.7 ROLE OF SPECTRAL REFLECTANCE CURVES IN REMOTE SENSING

The transmitted, emitted, absorbed, and reflected energy from the Earth features such as soil, vegetation, water, etc. at various wavelengths permits detection of them by the remote sensing sensors. The increased availability of remotely sensed data and the studies conducted with them have resulted in accumulation of reflectance, emittance, transmission, and absorption information about various features for more than three decades. Moreover, the detection properties of various features at different wavelengths are studied for a wide extent. The reader is recommended to refer to Rencz (1999) and Ustin (2004) for detailed description of spectral signatures and their use in remote sensing of various Earth features such as soil and related processes, forests, water resources, agricultural areas, various minerals, etc. In this section reflectance properties of common features and their detection ranges in the electromagnetic spectrum is explained.

The spectral reflectance curves given in Figure 2.9 are the mean reflectance curves of soil, water, vegetation, snow, and desert. For the same feature, the curves vary when the considered Earth feature has different properties. Although the soil curve in Figure 2.9 is smoother than the vegetation curve, it has troughs at 1.4, 1.9, and 2.2 μm. The troughs in spectral reflectance curve correspond to absorption. The main factors influencing the reflectance characteristics of soil are grain size and roughness of the soil, which cause variations in texture, iron oxide, organic matter, and moisture content (Lillesand and Kiefer, 1999). Moreover, these factors have complicated interrelations which help the understanding of resultant spectral reflectance for certain soil properties. For example, when soil is dry and coarse-grained, it will have lower reflectance than dry and fine grained soil (Figure 2.10). On the other hand, in the presence of water, courser grains in the soil allow water to drain more than the fine grains and thus low moisture content of coarse-grained soils have higher reflectance than fine grained soils with high moisture content. The soils containing iron oxides and high

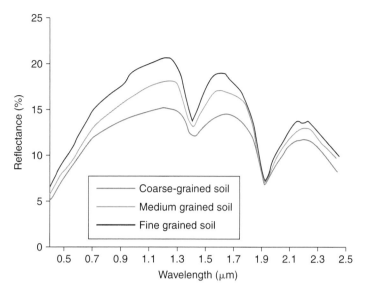

Figure 2.10 Spectral reflectance curves for typical dry soil with various grain sizes

degree of surface roughness have considerably lower reflectance than iron oxide-free and smooth soils in the visible portion of the electromagnetic spectrum.

The spectral reflectance curve for water, in Figure 2.9, shows that water generally has no or quite limited reflectance for wavelengths greater than $0.7\,\mu m$, which corresponds to the range of near infrared to thermal infrared or further parts of the electromagnetic spectrum. Clear water highly transmits energy in the blue and green wavelengths allowing investigation of underwater features by remote sensing. As near and thermal infrared wavelengths absorb most of the energy, detecting water bodies based on remote sensing relies mainly on working with these regions. However, such a perfect absorption characteristic is valid for deep and clean water bodies. If the water body is shallow or containing impurities such as chlorophyll, suspended matter, etc., there will be some amount of reflection either from the bottom part of the water or from the impurities in the water. Moreover, as the surface of water may behave like a mirror (especially for calm water bodies), causing large amount of specular reflectance, perfect absorption characteristics in the near infrared region may not be achieved. Figure 2.11 shows typical reflectance curves for snow, clear water, water containing sediments, and chlorophyll. When the water body contains some impurities such as sediments, chlorophyll, etc., its reflectance in visible and near infrared regions increases (Figure 2.11). The presence of chlorophyll in water results in absorption in blue range $(0.4–0.5\,\mu m)$, which corresponds to the first troughs in Figure 2.11, and increased reflectance in the green range $(0.5–0.6\,\mu m)$, which is the first peak in Figure 2.11. While the reflectance properties of water containing chlorophyll are mainly used for monitoring algae concentrations in water bodies, that for containing suspended matter provides information related to sediment transport to water bodies due to erosion. It is hardly possible to monitor directly other parameters associated with water quality

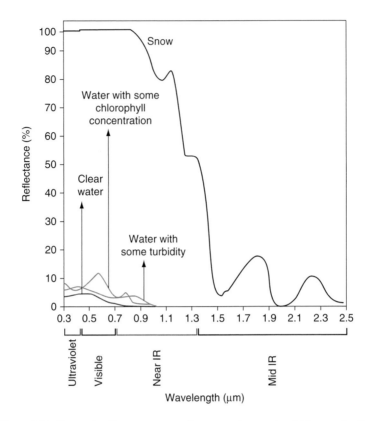

Figure 2.11 Spectral reflectance curves for snow, clear and turbid water bodies

such as dissolved oxygen, pH, etc. However, when these properties can be correlated to reflectance properties, water can also be monitored through remote sensing indirectly. In addition, water bodies like snow, glaciers, and clouds have almost perfect reflectance in visible wavelengths and have certain absorption characteristics in infrared wavelengths. These properties are used for detecting snow cover areas, monitoring glaciers, and clouds.

The vegetation reflectance curve given in Figure 2.9 characterizes certain properties related to vegetation which provide detection of vegetation cover and distinguishing vegetation types. There are three important factors affecting the reflectance properties of vegetation, namely, plant pigment, which is mostly correlated with the amount of chlorophyll, cell structure, and water content. Information related to plant pigments, cell structure, and water content of a vegetation cover can be obtained by investigating the reflectance behavior of vegetation in different wavelengths. Generally the visible wavelengths provide information about plant pigments (Figure 2.12). Information related to cell structure and water content can usually be obtained from near infrared and mid infrared wavelengths, respectively (Figure 2.12). The majority of electromagnetic energy is absorbed by chlorophyll in red and blue regions, while it is highly

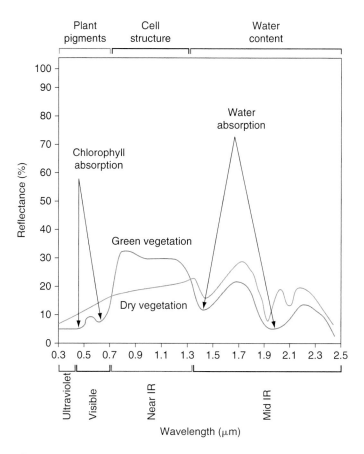

Figure 2.12 Spectral reflectance curves for green and dry vegetation

reflected in green. For this reason when leaves are healthy in summer, they are seen as green by our eyes. In fall, leaves become red, orange or yellow due to breakdown of chlorophyll (Verbyla, 1995). The high degree of absorption in blue and red regions and reflectance in the green region can also be observed as the first peak of the vegetation reflectance curve in (Figure 2.12). In the near infrared (NIR) region, most of the electromagnetic energy is reflected (between 30–50%) and the rest of it is transmitted with minimum amount of absorption (between 5–10%). The transmission of energy is mainly due to the high amount of air/cell interface area in the plants (Verbyla, 1995). This region allows the remote sensing analyst to detect three important properties of vegetation. First of all, as the reflectance property detected in the NIR wavelengths is related to cell structure, various vegetation types having different cell structures can be distinguished using the reflectance spectra in this region (Figure 2.12). Secondly, since stressed vegetation has a changed structure, the degree of stress can be detected in the NIR region. Finally, when the vegetation cover has more than one layer, the reflectance and transmission level of the energy increases due to multiple transmission and reflectance from various vegetation layers, which permits identifying the thickness of the vegetation canopy. The thicker the vegetation canopy, the less the reflectance in red wavelengths and the more the reflectance in NIR wavelengths. Thus, vegetation canopy is mainly detected by using indices based on the ratio of red reflectance values to NIR reflectance values. In mid infrared wavelengths there is quite limited transmission of the energy and most of the energy is either reflected or absorbed. The absorption in mid infrared wavelengths is due to the water content of the vegetation, and the main absorptions occur for the wavelengths of 1.4 and 1.9 μm (Figure 2.12). The reflectance, on the other hand, has the highest values in the mid infrared region for the wavelengths of 1.7 and 2.2 μm (Figure 2.12). When the water content in the vegetation decreases, like the dry vegetation reflectance curve in Figure 2.12, reflectance increases and the absorption decreases. Hence, the mid infrared region of the electromagnetic spectrum gives information about the water content of the vegetation as well as the health condition associated with the plants. A summary of reflectance, absorption and transmission behavior of water, soil, and vegetation for various wavelengths is given in Table 2.5.

2.8 REMOTE SENSING SYSTEMS

Remote sensing systems can be classified based on three perspectives (Figure 2.13). In perspective I, the sensor platforms are the main focus, in which the remote sensing is classified as ground-based, airborne-based, and space-borne-based systems. In perspective II, sensor systems are grouped into two categories as passive and active sensors. In perspective III, sensor systems are classified according to the sensor's ability to record spectral reflectance and spatial resolution.

The platforms on which the sensors can be located in perspective I is shown in Figure 2.14. Ground-based platforms are used when detailed information about very small areas on the Earth is required. They can provide highly precise and good quality data as they are least affected by the atmospheric interactions. The platform is usually located on the roof of a tall structure or on a crane. Airborne-based platforms are constructed by mounting remote sensing sensors to aircraft and used for collecting detailed ground information for larger areas. Processing airborne-based data requires more effort than the data from ground-based platforms as the data are affected by the flight and atmospheric conditions. They are also used for calibration of satellite-based remote sensing sensors. Remote sensing systems on space-borne satellites provide relatively good quality and inexpensive data as they obtain data for very large areas of the Earth. Space-borne-based platforms are mounted on satellites or spacecraft. With the advance of information technology, space-borne platforms are increasingly being used. The data from space-borne satellites require the highest degree of processing as they are influenced by various atmospheric effects and data streaming problems to the ground stations.

The classification of remote sensing systems in perspective II depends on the energy source radiating from the Earth. Passive sensors are designed to detect naturally occurring electromagnetic energy (Figure 2.15a). In passive sensor systems, the electromagnetic energy source is the Sun. The energy of the Sun is either reflected, as it is for visible wavelengths, or absorbed and then re-emitted, as it is for thermal infrared

Table 2.5 Basic Earth features and their spectral behavior under various electromagnetic wavelengths

Region		Wavelength Range (μm)	Basic Earth Features		
			Water	Soil	Vegetation
Visible	Blue	0.4–0.5	Water depth studies and underwater feature investigation due to high transmission of electromagnetic energy with much atmospheric scattering, identification of subsurface properties for water quality, coastal zone delineation	Soil type detection, identification of geological features	Vegetation type identification
	Green	0.5–0.6	Moderate transmission of electromagnetic energy with moderate atmospheric scattering, assessment of sediment and chlorophyll concentration	Almost none	Health and stress assessment
	Red	0.6–0.7	Water quality assessment with sediment and chlorophyll concentration, better identification of near surface properties of water due to limited transmission, least atmospheric scattering and absorption	Soil and geological boundary delineation	Type and plant growth condition
Infrared (IR)	Near IR	0.7–1.3	Assessing amount of suspended matter with turbidity identification with high reflectance level due to suspended particles	Soil moisture estimation	Type identification, crop growth monitoring, biomass estimation, vegetation health monitoring
	Mid IR	1.3–3.0	Distinguishing clouds from snow and ice, where high water content of snow and ice have lower reflectance than clouds	Soil moisture mapping	Crop monitoring
	Thermal IR	3.0–14.0	Water temperature predictions due to emissivity	Various Earth features that can be detected due to emissivity, features creating thermal (e.g. industrial areas, heat containing pipelines, geothermal sites, as well as detecting mineralogical content of rocks)	

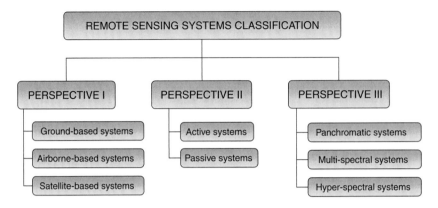

Figure 2.13 Classification of remote sensing systems with three different perspectives

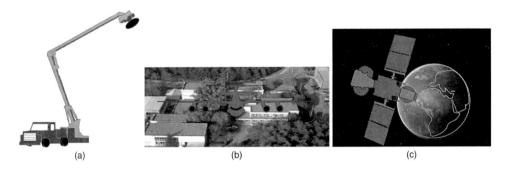

Figure 2.14 Types of sensor platforms: (a) ground-based platform, (b) airborne-based platform, and (c) satellite-borne platform

wavelengths. In remote sensing, many systems use passive detection. The active remote sensing systems provide their own energy source. In these systems, energy is directly sent to the target and received by the sensors in order to measure interaction of the target and the energy (Figure 2.15b). The sensor detects the reflection of the energy and measures the angle of reflection or the amount of time it takes for the energy to return. Active sensors require the generation of a fairly large amount of energy to adequately reach the targets. They can obtain measurements at anytime, regardless of the time of day or season. Active sensors can be used for examining energy types that are not sufficiently provided by the Sun such as microwaves by radar or sonar and laser by LIDAR (Light Detection And Ranging).

In perspective III, the classification of sensor systems are performed based on the spectral and spatial resolution of the sensors. The first type of sensors are called panchromatic/high resolution sensors in perspective three, where the sensors record electromagnetic energy ranging between 0.5–0.9 μm and the **spatial resolution** which is the smallest size of the object that can be detected by the sensor, is high. Multi-spectral sensors collect information in separate wavelength regions called **bands.** They usually record visible wavelengths (blue, green, and red) as well as near infrared wavelengths in

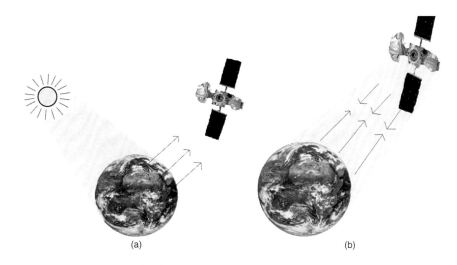

Figure 2.15 Passive (a) and active (b) sensor systems

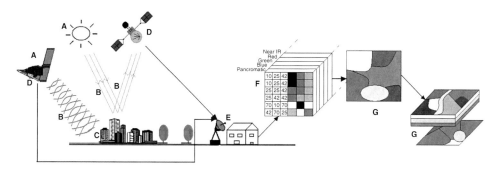

Figure 2.16 Components of remote sensing

four bands. Hyper-spectral sensors have ability to detect detailed spectral information by providing large number of bands with narrow spectral range. The details of sensor resolutions are given in Table A.1.

Regardless of the remote sensing classification perspectives, every remote sensing system consists of seven major elements (Figure 2.16). An energy source (A in Figure 2.16) is the most essential component for remote sensing to provide electromagnetic energy to the target of interest. Radiation and the atmosphere (B in Figure 2.16) provide the energy to travel from its source to the target and interact with the atmosphere it passes through. This interaction may take place twice as the energy travels from the target to the sensor. Interaction with the target (C in Figure 2.16) gives information related to the target and the radiation. Recording of energy by the sensor (D in Figure 2.16) is essential to collect and detect the electromagnetic radiation, after it is scattered, emitted, transmitted, reflected or absorbed. Transmission, reception, and processing (E in Figure 2.16) is needed for sending the energy recorded by the sensor, often in electronic form, to a receiving and processing station where the data are processed into an image. Interpretation and analysis (F in Figure 2.16) is required for obtaining remotely sensed information from image data, which can be done through visual and/or digital image analysis methods. Application (G in Figure 2.16) is the last component of the remote sensing process, which is basically conducted to reveal new information about the object of interest from the information extracted from the imagery for assisting in solving a particular problem. Once the data are obtained, it is processed and interpreted using remote sensing techniques which are extensively explained in Chapter 3.

REFERENCES

Aranof, S. (2005) *Remote Sensing for GIS Managers*. California, USA, ESRI Press.

Campbell, J.B. (2008) *Introduction to Remote Sensing*, Fourth Edition. New York, USA, The Guilford Press.

Cracknell, A.P. & Hayes, L. (2007) *Introduction to Remote Sensing*, Second Edition. Boca Raton, USA, CRC Press.

Fussel, J., Rundquist, D. & Harrington, J.A. (1986) On defining remote sensing. *Photogrammetric Engineering and Remote Sensing*, 52 (9), 1507–1511.

Holz, R.K. (1973) *The Surveillant Science Remote Sensing of the Environment*. Boston, Houghton Miffling Co.

Lillesand, T.M. & Kiefer, R.W. (2000) *Remote Sensing and Image Interpretation*. USA, John Wiley and Sons. Inc.

Milman, A.S. (1999) *Mathematical Principles of Remote Sensing*, Chelsea, MI, Sleeping Bear Press.

Rencz, A.N. (1999) Remote sensing for the earth sciences. *Manual of remote sensing*, Third Edition, Volume 3. USA, John Wiley and Sons.

Schowengerdt, R.A. (1997) *Remote Sensing Models and Methods for Image Processing*, Second Edition. San Diego, USA, Academic Press.

Ustin, S.L. (2004) Remote sensing for natural resource management and environmental monitoring. *Manual of remote sensing*, Third Edition, Volume 4. USA, John Wiley and Sons.

Verbyla, D.L. (1995) *Satellite Remote Sensing of Natural Resources*. USA, CRC Press LLC.

## Chapter 3

Remote sensing image analysis techniques

This chapter explains the nature and properties of remotely sensed data, its processing and interpretation. The image preprocessing methods like geometric and radiometric corrections, enhancement, and transformation techniques, which are essential prior to information extraction, are briefly described. The chapter then gives basic image interpretation methods used for mining applications. The principles of classification, change analysis, and Digital Elevation Model (DEM) extraction are explained.

3.1 CHARACTERISTICS OF REMOTELY SENSED DATA

The detected and stored electromagnetic energy forms the basis of remotely sensed data, which can be obtained photographically or electronically. When the data are obtained photographically, the electromagnetic energy is detected and recorded on photographic films which are sensitive to light. The data are recorded on the film material, which chemically reacts with the electromagnetic energy, in the form of light. Detecting electromagnetic energy electronically, on the other hand, relies on collecting generated electrical signals caused by the variations in the electromagnetic energy for a given view. The signals are later recorded on magnetic media. Therefore, by photography, the electromagnetic energy is detected and recorded at the same time on the photographic film and developing the film provides the remotely sensed **data in analog form**. The remotely sensed data obtained electronically are the **data in digital form**. Conversion between these formats is possible. Digital data can be recorded on a photographic film to obtain analog data, or analog data can be transformed to digital data by using scanners. In remote sensing, the term **image** refers to data in picture form. Thus, the term image covers both analog (photographic) and digital data. However, images are not always photographs (e.g. an image recorded by an infrared sensor is an infrared image but not infrared photograph). In general, analog data (i.e. photographic films/photographs) are easy and inexpensive to obtain, contain high spatial detail, and provide geometric integrity (Lillesand and Kiefer, 2000), whereas digital data can be obtained in various spectral resolutions and allow analysts for better storage and transmission capabilities.

A digital image consists of equal-sized picture elements called **pixels**. The electromagnetic energy reflected, emitted, etc. (radiance) from the Earth objects are recorded in each pixel in the form of digital numbers called **brightness values** or **digital numbers** (DN). Figure 3.1 illustrates a typical digital image. As can be seen from Figure 3.1, the image is in the form of a picture with some tonal variations and these tonal variations are represented by digital numbers in a two dimensional array.

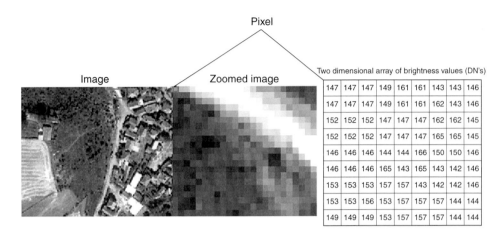

Figure 3.1 A typical digital remote sensing data

The pixels in the remotely sensed images contain two types of information, namely spatial and spectral. The pixel size provides spatial information from which corresponds to the covered area of the Earth's surface. The spectral response characteristics of the Earth features for a given electromagnetic spectrum range present the spectral information, which are recorded as brightness values. When one range of the electromagnetic spectrum is obtained as an image and visualized, such images yield tonal variations in the form of gray levels or black-white images, which are also called **panchromatic** images (Figure 3.1). Many remote sensing sensors have the ability of recording various ranges of the electromagnetic spectrum that yield a set of images. Such image sets are called **multispectral** images (Figure 3.2). Each image in the multispectral image set are also called **bands**. Although the mechanism of active remote sensing sensors, which usually work in the microwave region in the electromagnetic spectrum and have radar sensors (e.g. synthetic aperture radar, SAR), are different from passive remote sensing systems, which work in fine ranges of the electromagnetic spectrum and have optical sensors, the output of both systems are images with pixels containing tonal variations.

The remote sensing image data have four different resolutions, namely, spatial, spectral, radiometric, and temporal resolutions. The pixel size of remote sensing image determines the **spatial resolution** (Figure 3.3), in which a pixel is the smallest covered area on the Earth's surface with homogeneous tonal value. Hence, high spatial resolution refers to small pixel size containing higher detail of the Earth's surface, while low spatial resolution corresponds to large pixel size with lower detail of the Earth's surface. The spectral resolution defines the type of objects that can be detected by the remote sensing sensor, which is dependent on the object properties and its surroundings, image scale and sensor power as well as atmospheric and illumination conditions (Aranof 2005). When the object to be detected has contrast with respect to its surroundings and has a characteristic shape, it is more easily detected by the remote sensing sensor. Image scale is analogous to map scale, which is defined by the distance measured on the image divided by the same distance measured on the ground

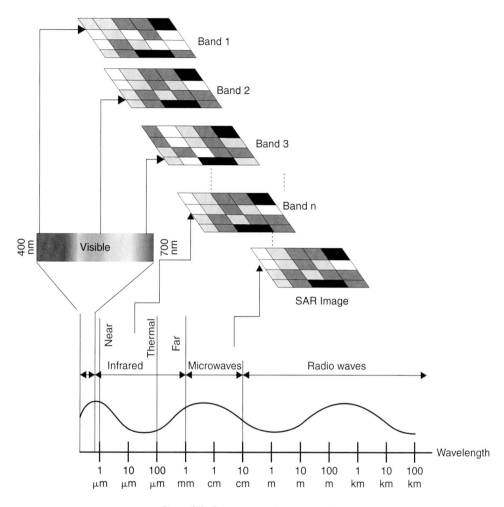

Figure 3.2 Remote sensing sensor data

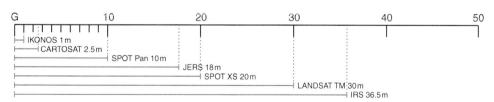

Figure 3.3 Spatial resolutions for various remote sensing sensor data

or ratio of focal length of the camera (distance between optical center of the lens and its focal point on the ground) to the flight altitude. For example, if an image is taken from a plane having altitude of 1000 m with a lens of 200 mm, the image will have a 1:5000 scale. Image scale affects the size of objects in the image, where larger objects are easily detected than the smaller ones. For the aerial photos image scale varies with

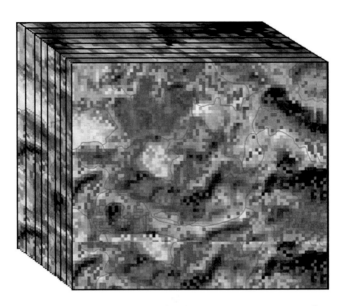

Figure 3.4 Spectral resolutions for the remote sensing sensor data

the flight altitude, however, satellite images have a fixed scale as they are mounted on space platforms. When atmospheric conditions like smoke, haze, etc. exist and/or illumination of the objects are not enough, the contrast in the images decreases. That affects the performance of the object detection processes.

The spectral details that are represented by the spectral bands determine the **spectral resolution**, where an image with high spectral resolution contains a large number of spectral bands with fine division of the electromagnetic spectrum. In other words, narrower division of the electromagnetic spectrum results in a larger number of bands, while wider division of it yields a lesser number of bands. Multispectral sensors typically have spectral resolutions ranging from 4 to 15 spectral bands. Hyperspectral sensors usually contain more than 200 spectral bands with very narrow divisions of the electromagnetic spectrum. Figure 3.4 shows a typical image cube that can be obtained from a hyperspectral sensor, where every layer in the cube represents an image of a narrow division of the electromagnetic spectrum. The range of brightness values or digital numbers (DN) in the image pixels denotes the **radiometric resolution**, which is expressed by the number of bits. The remote sensing sensor records DN in the form of binary series called **bits**. The n-bit binary image contains 2^n number of DN. For example an 8-bit image has DN ranging from 0 to 255 (2^8) whereas a 10-bit image have DN between 0 and 1023 (2^{10}). Thus the greater the bit sizes the more variable or finer detail of DN are contained in the image. Figure 3.5 illustrates the difference between the radiometric resolutions. Higher radiometric resolution provides better ability of separating the Earth's objects as they result in greater differences in brightness values (wider tonal variations). The frequency of the remote sensing sensor visiting the same geographical location and obtaining the images identifies the **temporal resolution**. The various resolutions related to commonly used images from satellite remote sensing sensors are given in Table A.1. The high temporal resolution

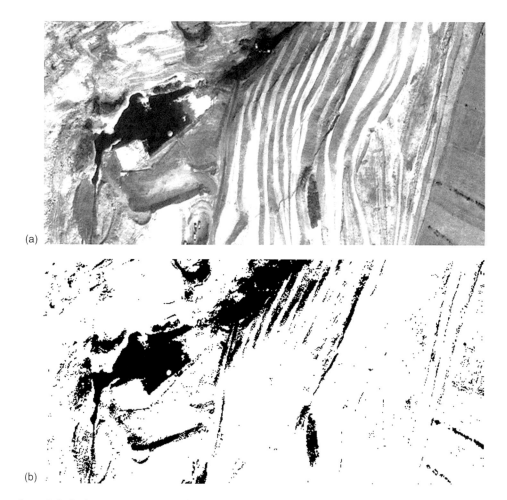

Figure 3.5 Radiometric resolutions for the remote sensing sensor data: (a) 16-bit image of a mine site and (b) 2-bit image of a mine site

brings about high frequency of obtaining images with shorter time between the successive visits of the same area (also called re-visit time). The temporal resolution determines the application area of the remote sensing information. High temporal resolution is mostly required for phenomena changing rapidly on the Earth like open pit mine environments, plant growth, water quality, etc. In this respect, satellite remote sensing systems are less costly in providing high spatial temporal data as compared to airborne systems.

The type of remote sensing application and required accuracy related to the application determines the optimum image resolutions. There are basically two types of accuracy in remote sensing, namely positional accuracy and feature accuracy. Positional accuracy refers to the object, location and size in the remote sensing image. Feature accuracy, on the other hand, is related to the certain features of the object like object type, shape, etc. Accuracy is assessed by comparing the sample set with true locations or features measured or collected on the ground and those obtained from the image. The comparison of these two sample set can be done through statistical methods or some accuracy indices, which will be explained in further sections. The positional and feature accuracies of the image data are directly related to suitability of the image for required applications. If the remote sensing application covers large areas, low positional and feature accuracies may be tolerated. High positional and feature accuracies are essential for remote sensing applications covering small areas. Positional accuracy of the remote sensing system is directly related to the spatial resolution, which is the function of camera/sensor, calibration, flight/orbital position, look direction of the camera/sensor, and quality of the sample set collected from the ground also called ground control points (GCPs). Feature accuracy is influenced by the spatial, spectral, radiometric, and temporal resolutions depending on the remote sensing analysis methods.

Due to the need for achieving the required accuracy in the remote sensing application, usually the obtained image data goes through certain image enhancements with respect to spatial, spectral, and radiometric aspects, called image pre-processing.

3.2 PRE-PROCESSING OF IMAGE DATA

Pre-processing refers to the processes which are applied to the images before starting extraction of information from the images called image interpretation. It can be performed mainly in three levels. The first level pre-processing is generally essential for most of the images and involves **geometric and radiometric correction**. The second level is mostly required for visual interpretation of the images as well as preparing the image for further digital image interpretation and provides improvement in the ability of separating the objects in the image, also called **image enhancement**. The third level of pre-processing is carried out in order to obtain several image characteristics based on transformation of images using measures of texture, vegetation indices, etc. and called **image transformation**. The need for pre-processing depends on the product type of the data providers and the nature of image interpretation.

3.2.1 Level I-geometric and radiometric correction

Each pixel of the image data corresponds to an area on the Earth's surface, which has certain positional information in the form of coordinates. Hence, the remote sensing image data should be **georeferenced** (i.e. images should contain geographic coordinates) in order to locate the images on the Earth. The process of providing geographic coordinates to each pixel of image data is called **georectification**. Usually, remote sensing sensors are coupled with inertial measurement systems (IMU) for continuous measurement of tilt and attitude of camera during image capturing and differential GPS (Global Positioning Systems), DGPS, for recording the camera position (i.e. X, Y, and Z coordinates) during image taking. Therefore, an image can provide each pixel's position on the Earth. However, such raw coordinates contain errors due to several factors, which result in geometric distortion of the image pixels. These geometric distortions have to be corrected prior to any image interpretation as the pixel geometries of the raw image do not reflect the actual object characteristics.

There are three main factors affecting the geometric distortions. During the image capture process, the camera parameters like attitude, tilt, and velocity of the platform may not be constant throughout the process causing geometric distortions. In optical remote sensing two types of camera: frame camera and line scanners, create different image geometries, which bring about different geometric distortions. The relief displacements on the ground are the third source of the geometric distortions. Without correcting the geometric distortions, an image interpretation cannot be used as a map product, since the coordinates of the extracted objects from the image do not correspond to their actual locations.

Geometric correction of images is a two-step procedure. In the first step, mathematical transformation is applied to the raw image coordinates to obtain actual locations of the pixels. During this transformation, the centre of the pixels in the original image changes which requires assignment of radiometric values (DN or brightness values) to the new pixel centers. The assignment of radiometric values to the transformed image pixels is called **resampling**.

The method of mathematical transformation of pixel coordinates depends on the available data and the required accuracy. In this procedure, the simplest and the most widely used approach is to find the parameters of a transformation function based on some known actual coordinates. The actual coordinates of known features are usually the ones which are visually recognized on the image like crossing roads, building corners, etc. from the ground (Figure 3.6). These known features with actual coordinates are called ground control points (GCPs). The actual coordinates of the objects can

Figure 3.6 GCP collections from the road crossings on the image (See color plate section)

either be obtained from direct measurement of the coordinates using DGPS devices on the ground or getting them from previous map products produced for the image location. When obtaining GCP coordinates from the map products accuracy in the map product should be considered. Usually the accuracy of maps is indicated by the data provider. Then the parameters of a transformation function are obtained using GCPs. Once the parameters are computed, they are applied to each image pixel to find the actual locations of them. The mathematical transformation function is applied for correcting X and Y coordinates of the pixels, not the height of the pixels. Hence, image rectification using GCPs uses 2D transformation functions. The main logic behind the use of transformation functions is finding the correlation between the image pixel coordinates and the actual ground coordinates. For this reason, the geometric correction based on GCPs does not consider errors due to camera parameters as well as relief displacement. Therefore, it is suitable only for the images acquired perfectly vertical to the ground (also called nadir viewing) with considerably flat terrain, which are free from relief effects.

The 2D transformation functions are in the form of polynomials with varying degree. The choice of polynomial degree depends on the required accuracy, available number GCPs and type and level of geometric distortions. The commonly used transformation functions are:

- Affine,
- Bilinear,
- Quadratic,
- Bi-quadratic,
- Cubic, and
- Bi-cubic.

Affine transformation provides a purely linear relation between the GCPs (X and Y) and the image coordinates (x and y) as given in Eq. 3.1

$$X = a_0 + a_1 x + a_2 y$$
$$Y = b_0 + b_1 x + b_2 y$$

(3.1)

The coefficients a_0, a_1, a_2, b_0, b_1, and b_2 should be calculated using GCPs. Since there are six coefficients to be calculated, at least three GCPs with their X and Y coordinates are required, theoretically. However, in practical applications more GCPs like 5–6 are required in order to evaluate the errors due to affine transformation. Affine transformation is suitable for the cases where geometric distortions are in the form of the ones illustrated in Figure 3.7.

Bilinear transformation involves additional polynomial terms of $a_3 xy$ and $b_3 xy$ (Eq. 3.2) and it corrects for the geometric distortions shown in Figures 3.7. As bilinear transformation needs the calculation of eight coefficients, theoretically the required number of GCPs should be at least four (four pairs of X and Y). Due to the need for error checking, practically the required number of GCPs is around 7–8, for bilinear transformation.

$$X = a_0 + a_1 x + a_2 y + a_3 xy$$
$$Y = b_0 + b_1 x + b_2 y + b_3 xy$$

(3.2)

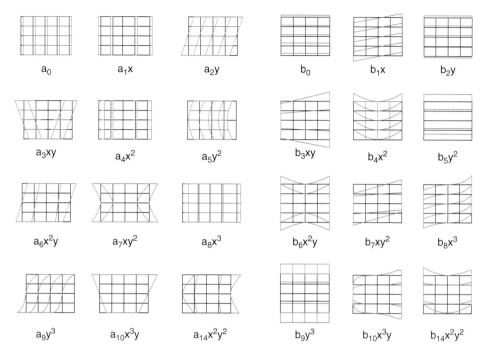

Figure 3.7 Transformation functions for geometric distortions/correction (Aronof, 2005)

The Quadratic transformation function is a second order polynomial with a total of 12 coefficients to be obtained from at least six GCPs (Eq. 3.3). Actually, this transformation requires 10–12 GCPs to evaluate the errors as well as the polynomial coefficients. It corrects for the set of geometrical distortions given in Figure 3.7.

$$X = a_0 + a_1x + a_2y + a_3xy + a_4x^2 + a_5y^2$$
$$Y = b_0 + b_1x + b_2y + b_3xy + b_4x^2 + b_5y^2$$

(3.3)

The Bi-quadratic transformation function has six more polynomial terms added to Eq. 3.3 (Eq. 3.4). The total number of polynomial coefficients to be calculated is 18, which requires at least nine GCPs. Bi-quadratic transformation practically needs GCPs between 16 and 18. It provides geometrical corrections for the distortions shown Figure 3.7.

$$X = a_0 + a_1x + a_2y + a_3xy + a_4x^2 + a_5y^2 + a_6x^2y + a_7xy^2 + a_8x^2y^2$$
$$Y = b_0 + b_1x + b_2y + b_3xy + b_4x^2 + b_5y^2 + b_6x^2y + b_7xy^2 + b_8x^2y^2$$

(3.4)

The Cubic transformation function contains third order polynomial terms with 20 coefficients to be evaluated from at least 10 GCPs (Eq. 3.5). The bi-cubic transformation on the other hand involves higher order terms requiring evaluation of 32 polynomial coefficients with minimum of 16 GCPs (Eq. 3.6). The required number of GCPs with quality checks raises these GCP requirements to 18–20 and 30–32 for cubic and bi-cubic transformation equations, respectively (Figure 3.7).

$$X = a_0 + a_1x + a_2y + a_3xy + a_4x^2 + a_5y^2 + a_6x^2y + a_7xy^2 + a_8x^3 + a_9y^3$$
$$Y = b_0 + b_1x + b_2y + b_3xy + b_4x^2 + b_5y^2 + b_6x^2y + b_7xy^2 + b_8x^3 + b_9y^3$$

$$(3.5)$$

$$\begin{aligned} X &= a_0 + a_1x + a_2y + a_3xy + a_4x^2 + a_5y^2 + a_6x^2y + a_7xy^2 + a_8x^2y^2 \\ &\quad + a_9x^3 + a_{10}y^3 + a_{11}x^3y + a_{12}xy^3 + a_{13}x^3y^3 + a_{14}x^3y^2 + a_{15}x^2y^3 \\ Y &= b_0 + b_1x + b_2y + b_3xy + b_4x^2 + b_5y^2 + b_6x^2y + b_7xy^2 + b_8x^2y^2 \\ &\quad + b_9x^3 + b_{10}y^3 + b_{11}x^3y + b_{12}xy^3 + b_{13}x^3y^3 + b_{14}x^3y^2 + b_{15}x^2y^3 \end{aligned}$$

$$(3.6)$$

Geometric distortions in the images can more precisely be corrected using so-called **sensor models,** which are transformation functions of camera parameters like orientation and position of the camera during image acquisition. The sensor models are coupled with a limited number of GCPs and give better accuracy than correction based on only GCPs. In order to perform geometric corrections based on sensor models, image data should contain sensor parameters or such data should be handled with special software products (e.g. PCI Geomatica, Erdas Imagine, Z/I, etc.) that have embedded sensor models. This type of geometric correction does not consider the distortions due to terrain relief.

In order to correct for geometrical distortions due to relief displacements, a digital elevation model (DEM) for the Earth's surface over which the image is obtained, should be included in the transformation functions. The DEM for the image can either be obtained from other data sources or from processing of stereo image products, which will be explained in further sections. Correction for relief displacements has a critical role for the images taken from mountainous terrain as well as high-resolution images. Geometric correction of the image for camera tilt and attitude as well as the relief displacements using DEM data is called **orthorectification.** In orthorectification, an elevation value for each pixel of the image is obtained from DEM data and incorporated into transformation functions, which are based on sensor models. In this way, the image is corrected for all types of errors like sensor and relief. Orthorectified images can be used as map products since geometric distortions in the image are precisely corrected.

Correcting the geometric distortions of the image, in fact, creates a new set of image pixels representing the actual locations of pixels on the Earth. However, this new corrected pixel set does not contain radiometric (DN/brightness) values, which should be assigned by **resampling.** As the original pixel geometry changes after applying transformation functions yielding a **rectified image,** the DN of the corrected image can be obtained from the neighbors of the corrected pixel (Figure 3.8). There are basically three methods for resampling:

- The nearest neighbor,
- Bilinear interpolation, and
- Cubic convolution.

In the **nearest neighbor** resampling method, the center coordinate of each pixel of the rectified image is found in the original image (raw image before rectification) and the nearest pixel center's DN in the original image is assigned to the rectified

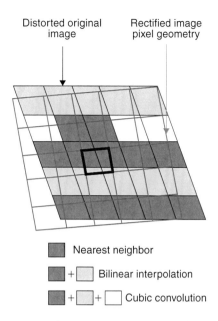

Distorted original image

Rectified image pixel geometry

■ Nearest neighbor

■ + □ Bilinear interpolation

■ + □ + □ Cubic convolution

Figure 3.8 Resampling strategies for geometric corrections (modified from Aronof, 2005)

image pixel (Figure 3.8). In the **bilinear interpolation** strategy, the four nearest neighbor pixel's DN are considered. The weighted average of the four neighboring DN's is calculated and assigned to the pixel (Figure 3.8). In the **cubic convolution** method, the weighted average of the sixteen neighboring pixels based on the distance to the rectified pixel centre is evaluated and assigned to the pixel of the rectified image (Figure 3.8). As the nearest neighbor method assigns the original DN's to the rectified image, this sampling method is mostly suitable for image interpretations which require original radiometric values like land use land cover classification. However, this sampling method brings about the multiple or non use of the same DN's that yields pixel blocks in visualization. Hence, the nearest neighbor resampling method is not well suited to visual interpretation of the image. On the other hand, bilinear interpolation and cubic convolution resampling methods yield smoother images due to obtaining DN's through averaging. The most visually interpretable results are obtained from cubic convolution as it gives the highest amount of smoothing.

Radiometric correction refers to the compensation and removal of errors in the radiometric values (DN/brightness values). The errors in the DN arise from sensors and atmospheric effects. Usually the sensor errors occur due to improper functioning of the sensor, which are eliminated in most of the recent sensors and exist mostly in remote sensing image data obtained from old sensors like Landsat image data (Table A.1). The most widely faced error in such sensor data is de-stripping that is compensated by assigning the radiometric values of the neighboring pixels to the erroneous pixels. The other types of errors are lines of pixels containing no DN (missing lines) and pixels containing wrong DN's, which appear as salt-and-pepper like noise. Although the earliest remote sensing sensors are not now in use, the data provided by such sensors are still valuable since they have a role in monitoring the long term changes as a reference data set. The recent developments in the sensor technology supply data free from improper sensor functioning. However, atmospheric effects resulting in reduction in image quality still exist in current image products. The atmospheric effects usually change the nature of electromagnetic energy as discussed in Chapter 2, resulting in decreased contrast in the images. Atmospheric correction algorithms require rigorous models involving measurements of atmospheric conditions on the date and time of image acquisition. These algorithms are mostly embedded in remote sensing software products. Atmospheric correction is essential for change detection algorithms as this image interpretation relies on comparing images of different date and time with respect to radiometric values. The contrast problems due to atmospheric effects can also be enhanced by image enhancement algorithms, which will be explained in further sections.

3.2.2 Level II-image enhancement

Unlike the level I type of pre-processing, level II type of image pre-processing is not compulsory prior to every image interpretation. However, it may become essential if some improvements in the image data are necessary for image interpretations. Visual image interpretation especially may require enhancement of spectral and spatial features. As image enhancement changes the original characteristics of the image data, there can be some information loss. Such loss of information may not be tolerated by certain image interpretation methods based on digital image analysis like classification that yields better interpretation performance without image enhancement.

The **spectral image enhancement** improves visual quality of the image by increasing the contrast and variability in brightness values between the pixels in the image data, which are quite poor in the raw image, especially for images from satellite sensors. This is due to the fact that sensors are designed for a wide range of spectral reflectance values for detecting a large variety of materials on the Earth, whereas the range of reflectance values for an image scene is relatively short resulting in similar brightness values for the majority of the pixels in the image data with low contrast.

An image histogram is one of the effective tools for assessing the radiometric quality of the images. It is obtained by grouping the brightness values of an image band into certain histogram classes (also called bins), and then plotting the number of pixels in each histogram class (frequency of brightness values) with respect to the classes. Hence, the image histogram reflects the distribution of brightness values in a certain image band. For images with poor contrast and brightness, the frequency plot (histogram) of DN for a given image band yields accumulation of certain DN's in a limited number of bins (histogram classes) as given in Figure 3.9. The histogram of the DN's for a raw image can also be shifted to one end of the histogram that results in relatively darker (values closer to zero, Figure 3.9a) or brighter images (values closer to the highest radiometric number depending on the radiometric resolution, Figure 3.9b). An image having a wide range of brightness values exhibits good contrast with better visual interpretability. The histogram of such images contains the full range of available radiometric values.

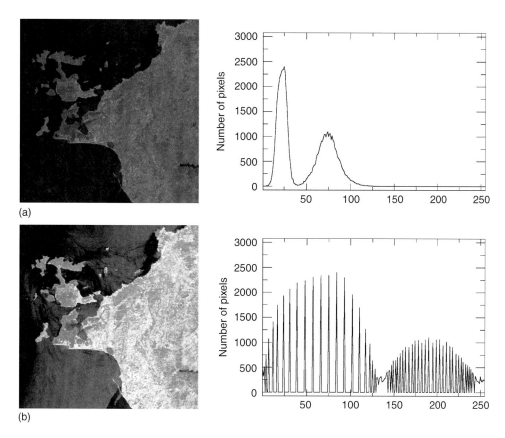

Figure 3.9 Histogram of brightness values (DN's) for a typical raw image data: a) raw image, b) enhanced image

The problem of poor radiometric quality is overcome using either linear or non-linear contrast enhancement methods depending on the distribution of the DN's of the image. If the distribution of DN's exhibits Gaussian (Normal) distribution or distributions closer to Gaussian with single mode (a bell-shaped histogram with only one peak) **linear contrast enhancement** can be applied. Linear contrast enhancement is also called histogram stretching, where the DN's of the original image are transformed to a new set of DN's using a transformation function. In this way a new set of DN's in the output image comprise the full range of radiometric values. A typical transformation function for histogram stretching is given in Eq. 3.7

$$DN_{out} = \left(\frac{DN_{in} - DN_{min}}{DN_{max} - DN_{min}} \right) \cdot 2^{n} \tag{3.7}$$

In Eq. 3.7:

DN_{out} : DN of a pixel after histogram stretching (output image)
DN_{in} : DN of a pixel before histogram stretching (input image)

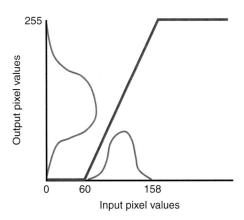

Figure 3.10 A typical LUT for linear contrast enhancement

DN_{min} : The minimum DN in the input image
DN_{max} : The maximum DN in the input image
n   : Bit level of image $(0, 1, 2, \ldots, 64)$

For example if the DN's (DN_{in}) in the red band of an 8-bit (n = 8) image (DN_{in}) range between 55 (DN_{min}) and 165 (DN_{max}), after applying Eq. 3.7 the DN's will range between 0 to 255. Eq. 3.7 also allows one to establish a table for one to one correspondence between DN's of the original and the output images. A **Look-up-table (LUT)** serves for this purpose. Most of the remote sensing software products provide LUT after image enhancement (Figure 3.10).

The linear contrast enhancement provides an image with the full range of radiometric values causing pixels to be visually more separable (Figure 3.11). Nevertheless, it creates the problem of unequally enhancing the features in the image. This problem may cause diminishing of certain salience pixels that may have a critical role in image interpretation. The problem can be partially solved if certain Earth features are of concern. Eq. 3.7 can only be used for the considered features instead of taking the whole image into account. For example, if the purpose is to monitor vegetation health for a rehabilitated mine site, the range of DN's for the considered vegetation in the red band is used in Eq. 3.7 instead of whole range of DN's in the image. This type of contrast enhancement is also known as **piece-wise contrast enhancement**. The **non-linear contrast enhancement**, also called **histogram equalization**, would be a solution to the shortcoming of histogram stretching. In histogram equalization DN's are enhanced based on their relative frequencies. Hence, more frequently occurring DN's are enhanced more than the others (Figure 3.11). The histogram equalization is usually suitable for non-Gaussian DN distributions as well as bi-modal distributions.

The contrast enhancement methods provide global improvement of spectral properties of an image, where the enhancement techniques are based on global image parameters like minimum and maximum of DN's. It is also possible to enhance images using local image parameters, which is called **spatial filtering** or **edge enhancement**.

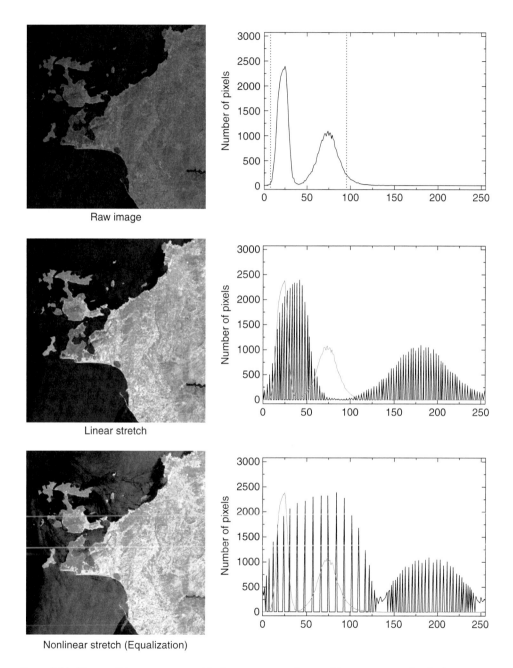

Raw image

Linear stretch

Nonlinear stretch (Equalization)

Figure 3.11 Linear (histogram stretching) versus non-linear (histogram equalization) contrast enhancement

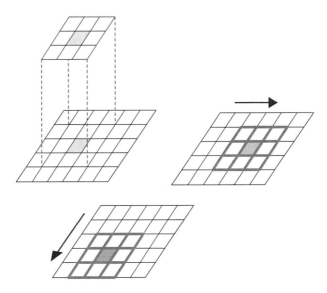

Figure 3.12 Movement of a 3 × 3 kernel on the image (modified from Lillesand and Kiefer, 2000)

The spatial filtering relies on moving window approach. A moving window (also called kernel) is a local operator with a size that defines the neighborhood of a given pixel (Figure 3.12). The DN value of a pixel in the centre of the window (kernel) is enhanced using arithmetical operators for various statistics or coefficients of neighboring pixel's DN's around the centre. For this reason the kernel size constitute odd numbers like 3 × 3, 5 × 5, 7 × 7, 9 × 9 providing a centre pixel with its neighbors. This procedure is also called **convolution**, which is an image processing term. An image convolution basically has four steps:

1   A kernel size (window size) is determined.
2   An array corresponding to the kernel is established by filling the array entries with weighting factors.
3   The kernel center is set to a pixel of the image and all the pixels within the kernel is multiplied by the corresponding array weights. Then the sum of the multiplication is assigned to the centre pixel.
4   The output image is obtained by repeating Step 3 for each pixel of the image.

Therefore, the enhancement is performed based on local DN's with a given window size. When the window is moved throughout the whole image, every pixel has locally enhanced DN's. In this approach, the size of window determines the level of local enhancement. Large window size makes the enhancement approach to global enhancement. The window size selection depends on the size of the image and that of the target features to be extracted from the image. The spatial filtering using moving windows results in enhancement of the edges in the image. For this reason such local enhancement is also called as **edge enhancement**. The edges in an image are resultant

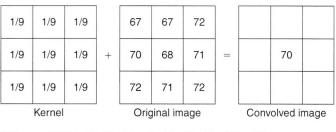

$$DN_{out} = 1/9*(67+67+72+70+68+71+72+71+72) = 70$$

Figure 3.13 Mean filter implementation

from sudden changes in the DN's. The spatial filtering sharpens or smoothes the pixels having sudden changes yielding an image with increased contrast around the edges due to moving window approach. The nature of spatial filtering depends on the type of edges to be enhanced. There are basically two types of spatial filters:

- Low-pass filters
- High-pass filters

Low-pass filters smooth the images by reducing the high spatial frequency components (edges). Therefore, large areas containing small edge details are blurred and thus become more homogeneous and easily detectable. This type of filters are usually employed reducing salt-and-pepper type of noise as well as eliminating details in the image to highlight boundaries of large homogeneous regions such as forest, agriculture, dump site, surface mine layout, etc. The degree of smoothing in a low-pass filter operation is controlled by the window size. The larger the window size the greater is the smoothing. The low-pass filter basically employs a convolution operator for a given window (kernel) considering the central tendency measures of the DN's in the kernel such as mean, median, and mode. In other words, one of the central tendency measures are evaluated for the DN's in the kernel and the computed value is assigned to the centre pixel of the kernel (Figures 3.13 and 3.14). Figures 3.15 and 3.16 illustrate the results of mean and median filters with 11 × 11 kernel sizes. The effect of using various kernel sizes is illustrated in Figure 3.17, where the largest kernel size yields the highest blurred image.

In contrast to low-pass filters, the **high-pass filters** sharpen the edges that are high spatial frequency components of the image. The high-pass filters are mostly used for detecting linear features with various scales ranging from geological structures (e.g. faults, lithological boundaries) to man-made structures (e.g. buildings, roads). There are various types of high-pass filters for different edge enhancement purposes. Usually linear high-pass filters are suitable for enhancing highly linear features such as faults, highways, etc., while the non-linear high-pass filters are mostly appropriate for highlighting polygonal features like buildings. A typical set of 3 × 3 convolution operators to be implemented for directional linear edge detection filters are given in

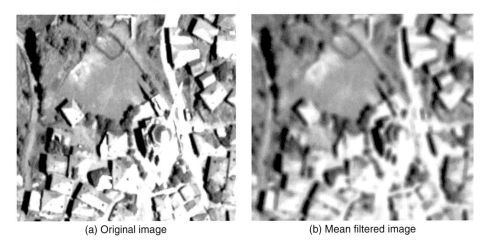

67	67	72
70	68	71
72	71	72

=

	71	

Original image   Convolved image

DN_{in} = 67, 67, 72, 70, 68, 71, 72, 71, 72

DN_{rank} = 67, 67, 68, 70, _71_, 71, 72, 72, 72

DN_{out} = 71

Figure 3.14 Median filter implementation

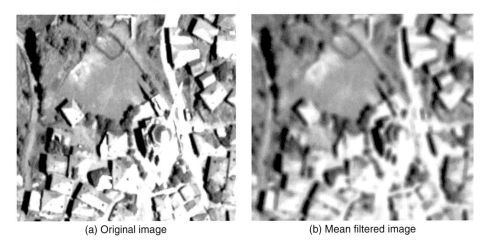

(a) Original image   (b) Mean filtered image

Figure 3.15 Mean filter applied to an image with 11 × 11 kernel size

Figure 3.18. As can be seen from Figure 3.19, edges in vertical, horizontal, and diagonal directions are enhanced by employing respective convolution operators.

Figure 3.20 illustrates one of the most widely used non-linear high-pass filters called **Sobel**. The DN of a pixel for Sobel is computed using Eq. 3.8

$$DN_{out} = \sqrt{X^2 + Y^2}$$
$$X = (DN_3 + 2DN_6 + DN_9) - (DN_1 + 2DN_4 + DN_7) \qquad (3.8)$$
$$Y = (DN_1 + 2DN_2 + DN_3) - (DN_7 + 2DN_8 + DN_9)$$

In Eq. 3.8:

DN_{out} : DN of a pixel in the centre of the kernel (output image)
DN_i : DN of the ith pixel in the kernel

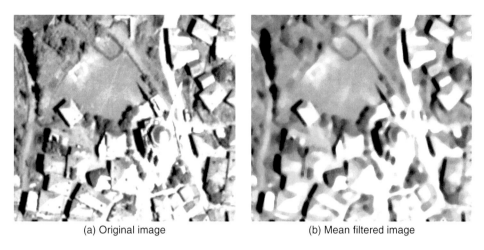

(a) Original image   (b) Mean filtered image

Figure 3.16 Median filter applied to an image with 11 × 11 kernel size

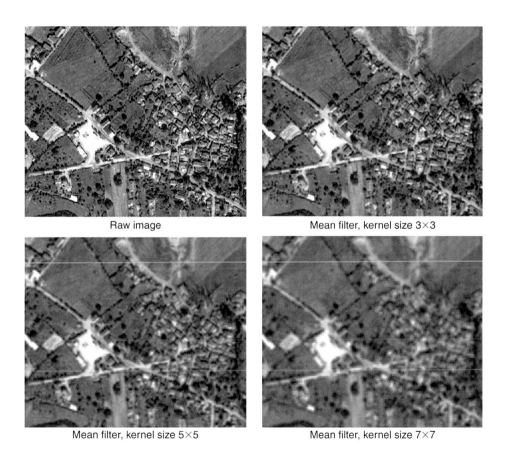

Raw image   Mean filter, kernel size 3×3

Mean filter, kernel size 5×5   Mean filter, kernel size 7×7

Figure 3.17 Effect of filter size in filtering

Linear edge detection

Vertical			Horizontal			Diagonal					

−1	0	1
−1	0	1
−1	0	1

N-S

−1	−1	−1
0	0	0
1	1	1

E-W

0	1	1
−1	0	1
−1	−1	0

NW-SE

1	1	0
1	0	−1
0	−1	−1

NE-SW

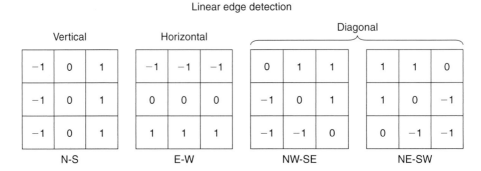

Figure 3.18 Linear high-pass filter kernels

(a) Original image   (b) Linear high-pass filtered image

Figure 3.19 Linear high-pass filtering

3.2.3 Level III-image transformation

Transforming images provides additional information for image interpretations both using digital image analysis and visual techniques. The transformed images can serve as either direct interpretation results or intermediate results to be used for further processing. Image transformation methods can be divided into two categories:

- Transformations using a single band of the image
- Transformations using several bands of the image

The most frequently applied image transformation to a single band of the image is extraction of texture measures. Similar to spatial filtering, texture measures are local features inherent in the image. Texture in the image is resultant from the degree of variability in the DN's. The highly variable DN's of a certain part of the image causes

BV$_1$	BV$_2$	BV$_3$
BV$_4$	BV$_5$	BV$_6$
BV$_7$	BV$_8$	BV$_9$

$$Sobel_{out} = (X^2 + Y^2)^{1/2}$$

$$X = (BV_3 + 2BV_6 + BV_9) - (BV_1 + 2BV_4 + BV_7)$$

$$Y = (BV_1 + 2BV_2 + BV_3) - (BV_7 + 2BV_8 + BV_9)$$

−1	0	1
−2	0	2
−1	0	1

X

1	2	1
0	0	1
−1	−2	−1

Y

Figure 3.20 Non-linear high-pass filter kernel for Sobel

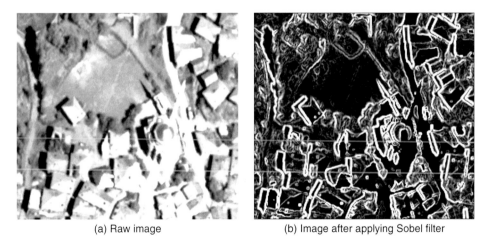

(a) Raw image   (b) Image after applying Sobel filter

Figure 3.21 Sobel filter application

a rough texture, while relatively small variability brings about smooth texture indicating homogeneity. Texture measures play an important role in digital image interpretation, especially for images with high spatial resolution as they help in extracting inherent properties from the image data. The most frequently used texture measures are gray level co-occurrence matrix (GLCM) obtained based on various statistics evaluated for a given kernel. The detailed description of texture measures can be found

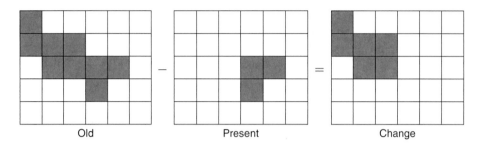

Figure 3.22 Transformation by subtraction

from Haralick *et al.* (1973), Jain (1989), Pratt (1991), Gonzalez and Woods (1992), and Schowengardt (1997).

There are various transformations based on multiple bands on the image data. The most frequently used ones are:

- Transformation based on arithmetical operations,
- Vegetation indices,
- Principal component analysis (PCA), and
- Transformation into different color spaces.

Transformation based on arithmetical operations is either applied for the same bands of images obtained for different dates or different bands of the same image. Subtraction and division are two arithmetic operators used in the transformation. Subtraction is mostly used for identifying the changes in the images obtained from two different dates. For example if two images are obtained for an area in the past and the present (Figure 3.22) with DN's of DN_{old} and $DN_{present}$ for a given band, the output image obtained by $DN_{old} - DN_{present}$ will show the changed areas. As the substraction may yield negative values usually the output image is obtained by adding a constant (2^n) to $DN_{old} - DN_{present}$ expression, where n is the bit number of the image. Therefore, in the output image the pixels with values close to the constant indicate the unchanged areas. The pixels with DN's greater or lower than the constant constitute the changed areas.

Instead of subtraction, images with two different dates for a given area can be divided. In this case, the DN's close to 1 in the resultant ratio image show the unchanged areas. However, the division operator is mostly used for different spectral bands of the same image, which is also called **spectral ratioing** (Lillesand and Kiefer, 2000). Finding the ratio of two different spectral bands (e.g. DN_{red}/DN_{blue}) allows the analyst to extract features which are difficult to discriminate as the ratio image highlights such features. Moreover, a ratio image reduces the illumination effects caused by the topography and hence provides better discriminating properties. For example, the same forest cover on two sides of a hill with one side illuminated and the other shaded, may have different DN value range, which may result in different vegetation classes. In this case, taking the ratio of spectral bands yields the same ratio for the same vegetation cover. Figure 3.23 shows the ratio images of two different time periods for a mine site

Red   Green   Blue

Figure 3.23 Transformation by division (See color plate section)

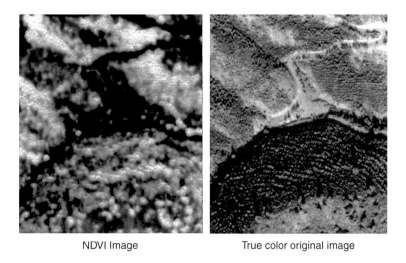

NDVI Image   True color original image

Figure 3.24 NDVI for a surface coal mine (See color plate section)

applied to Red, Green, and Blue bands. Colored pixels in Figure 3.23 show the changed pixels.

Vegetation indices are the index values obtained from the spectral bands of the image for identifying the vegetated areas (Figure 3.24). Although there are numerous indices introduced in the literature (Table 3.1), the most widely used vegetation index is the **normalized difference vegetation index (NDVI)**. NDVI relies on subtraction and division operations using red and near-infrared bands as given in Eq. 3.9.

$$DN_{NDVI} = \frac{DN_{NIR} - DN_{RED}}{DN_{NIR} + DN_{RED}} \tag{3.9}$$

In Eq. 3.9:

DN_{NDVI} : DN for NDVI image (output image)
DN_{NIR} : DN for NIR band
DN_{RED} : DN for RED band

An example of an NDVI image is illustrated in Figure 3.24.

A review of various types of vegetation indices is given by Glenn (2008). Table 3.1 summarizes the existing indices for extracting vegatation boundaries in the remotely sensed images.

Principal component analysis (PCA) is a statistical technique for discovering the pattern and reducing redundancy in a large multi-dimensional data like multi-spectral image data. Thus PCA is used for remote sensing image data either by eliminating the correlation between the bands, which is quite high in multi-spectral images, or discovering hidden patterns in such data. The result of PCA is several component images that can be used in both visual interpretations and digital image analysis based interpretations. PCA transforms the multi-dimensional data into a new space, where the correlation between the data set is eliminated with minimum loss of information in the data. If the image data have n number of bands, after applying PCA, the image has a lesser number of bands than n and it carries most of its information content. The transformation of original image data using PCA produces component images. The first principal component (PC1) carries the highest information, where 85–95% of the variance in the data is represented in PC1. The second (PC2), third (PC3), and other components contain less information components. When remote sensing analysis requires reduction in the dimensionality of the data, PC1 can be used as it carries most of the information with less data, which provides better computational efficiency for digital image analysis algorithms. On the other hand, if the analyses are focused on identification of change between multi-temporal image data or discovery of certain features, which are obscured by the majority of the data for a single imagery, then PC2, PC3 or other components will provide valuable information, as they reflect the remaining information after PC1. In two-dimensional space, a trend observed by plotting DN's of the first and the second band diminishes in the transformed data plotted with respect to PC1 and PC2 (Figure 3.25). Each principal component is expressed as the linear combination of n bands of the original image (Eq. 3.10)

$$DN_{PCi} = a_{i1}DN_1 + a_{i2}DN_2 + \cdots + a_{in}DN_n \tag{3.10}$$

In Eq. 3.10:

DN_{PCi} : DN for the ith component of the transformed image
DN_i : DN for the ith band of the original image, $i = 1, 2, \ldots, n$
a_{in} : Component weights obtained for each band and principal component from PCA

Transformation into different color spaces involves using different color spaces than red, green, and blue (RGB). The multispectral images are usually visualized by adding DN's in the bands of RGB. Such visualizations are in fact color composites of the image and visualizing the RGB color composite is called **true-color** as the actual

Table 3.1 Summary of vegetation indexes (After Gong et al., 2003)

Index	Equation	Definition	Reference
SR	$DN_{SR} = \dfrac{DN_{NIR}}{DN_{RED}}$	Simple Ratio	Baret and Guyot, (1991); Tucker, (1979).
NDVI	$DN_{NDVI} = \dfrac{DN_{NIR} - DN_{RED}}{DN_{NIR} + DN_{RED}}$	Normalized Difference Vegetation Index	Fassnacht et al. (1997); Smith et al. (1991).
PVI	$DN_{PVI} = \dfrac{1}{\sqrt{a^2 + 1}}(DN_{NIR} - aDN_{RED} - b)$ where a: slope of soil line b: soil line intercept	Perpendicular Vegetation Index (Mainly used for vegetation in arid/semiarid areas)	Baret and Guyot (1991); Huete et al. (1985).
SAVI	$DN_{SAVI} = \dfrac{(DN_{NIR} - DN_{RED})(1 + L)}{(DN_{NIR} + DN_{RED} + L)}$ where L: correction factor for soil, which is 0 for dense vegetation and can be 0.5 for sparse vegetation. L is correcting for reflectance from soil under the vegetation	Soil Adjusted Vegetation Index	Huete (1988); van Leeuwen and Huete (1996).
NLI	$DN_{NLI} = \dfrac{(DN_{NIR}^2 - DN_{RED})}{(DN_{NIR}^2 + DN_{RED})}$	Non-linear Vegetation Index	Goel and Qi (1994).
RDVI	$DN_{RDVI} = \dfrac{(DN_{NIR} - DN_{RED})}{\sqrt{(DN_{NIR} + DN_{RED})}}$	Re-normalized Difference Vegetation Index	Roujean and Breon (1995).
MSR	$DN_{MSR} = \dfrac{(DN_{NIR}/DN_{RED} - 1)}{\sqrt{(DN_{NIR}/DN_{RED})} + 1}$	Modified Simple Ratio	Chen (1996).
WDVI	$DN_{WDVI} = DN_{NIR} - aDN_{RED}$ where a: slope of soil line	Weighted Difference Vegetation Index	Clevers (1988); Clevers (1991).
MNLVI	$DN_{MNLVI} = \dfrac{(DN_{NIR}^2 - DN_{RED})(1 + L)}{(DN_{NIR}^2 + DN_{RED} + L)}$ where L: correction factor for soil, which is 0 for dense vegetation and can be 0.5 for sparse vegetation. L is correcting for reflectance from soil under the vegetation	Modified Non-linear Vegetation Index	Tarabalka et al. (2010).
NDVI*SR	$DN_{NDVI*SR} = \dfrac{(DN_{NIR}^2 - DN_{RED})}{(DN_{NIR} + DN_{RED}^2)}$	Combined SR-NDVI	Tarabalka et al. (2010).
SAVI*SR	$DN_{SAVI*SR} = \dfrac{(DN_{NIR}^2 - DN_{RED})}{(DN_{NIR} + DN_{RED} + L)DN_{RED}}$	Combined SAVI -SR	Tarabalka et al. (2010).
TSAVI	$DN_{TSAVI} = \dfrac{a(DN_{NIR} - aDN_{RED} - b)}{[aDN_{NIR} + DN_{RED} - ab + X(1 + a^2)]}$ where a: slope of soil line b: soil line intercept X: adjustment factor to minimize noise due to soil	Transformed Soil Adjusted Vegetation Index	Baret and Guyot (1991).

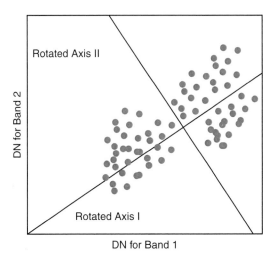

Figure 3.25 Graphical representation of original data and the transformed data in PCA

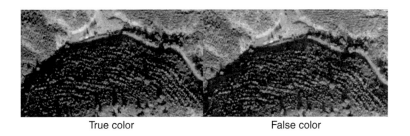

True color   False color

Figure 3.26 True and false color composites (See color plate section)

color of the features on the Earth is seen. It is also possible to visualize several band combinations in various color composites. One of the most frequently used color composite is called **false-color or color infrared,** where the near-infrared (NIR), red (R), and green (G) is visualized by adding the three bands. In false color images (NIR-R-G) red color represents vegetation and dark colors (black or near-black) represents water bodies (Figure 3.26).

The DN's representing RGB bands form the so-called color space as shown in Figure 3.26. Any color in the color composite from RGB bands can be represented in 3D coordinate system (RGB) (Figure 3.27), where the origin is black with value of zero. This representation in RGB coordinate axis is also called **color cube** with eight corners of red, green, blue, cyan, magenta, yellow, black, and white. Colors can be represented in different spaces other than RGB space. In digital image processing there are various color spaces. For a comprehensive discussion of color spaces see Gonzalez and Woods (1992). Among various color spaces, intensity-hue-saturation (IHS) color space is one of the most widely used color spaces as alternative to RGB space for remote sensing image data. IHS color space is illustrated by **color sphere** (Figure 3.28), where intensity

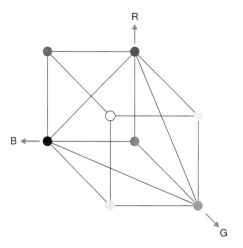

Figure 3.27 RGB color space in graphical form (See color plate section)

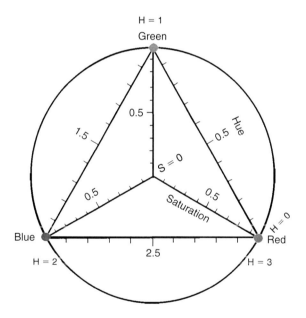

Figure 3.28 IHS color space graphical form (modified from Sabins, 1987) (See color plate section)

is on the vertical axis, hue is on the circumference and saturation is on the radial axis. Intensity (I) reflects the variability in the brightness ranging between black and white (0 (black)–255 (white) in 8-bit image). The dominant color in the wavelength is expressed by hue (H), which starts from 0 corresponding to red and increases counter-clockwise around the circumference of the sphere reaching 255 in 8-bit image. Finally, saturation (S) denotes the level of purity in color changing between 0 (the centre of the sphere) and 255 in 8-bit image (at the surface of the sphere). Thus S value of 0 corresponds to equal representation of all wavelengths, i.e. impure colors of gray. An S value of 255 in 8-bit image, on the other hand, represents pure colors with intense tones, while the S values between 0 and 255 illustrate pastel tones. The transformation of image data to IHS color space has two major roles in preprocessing. The first one has uses in image enhancement where independent contrast stretching can be applicable to one of the IHS bands, while leaving the rest of the dimensions unaffected. IHS transformation is also used for visualizing the geometrically corrected (georectified) images having different spatial resolutions or different remote sensing sensors.

3.3 INTERPRETATION OF IMAGE DATA

In order to extract information from the image data obtained from remote sensing sensors, images are required to be interpreted. Although there are numerous types of information that can be obtained from remote sensing image data, the interpretation methods frequently used for mining environments are considered in this chapter. Acquiring information from a mining environment mainly involves extracting the boundaries of features for the mining environment in the form of points, lines, and/or areas as well as obtaining the height of features, identifying the land use and land cover characteristics related to the mining environment and investigating the changes in the mining environment.

There are basically two types of interpretation namely, visual interpretation and interpretation based on image analysis methods. For the images with very high spatial resolution the human eye is an excellent interpreter. However, images obtained for wavelengths that are different from the visible range of the electromagnetic spectrum are difficult to interpret by the human eye. Moreover, interpreting the images by human eye requires skilled personnel who have appropriate experience. Image interpretation based on image analysis methods, on the other hand, can be performed with less experienced users and allows one to extract information from every band of the electromagnetic spectrum as well as providing automatic and/or semi-automatic interpretation. However, it requires execution of sophisticated algorithms and remote sensing software as well as some level of human interaction.

3.3.1 Visual image interpretation

Visual interpretation of remote sensing image data constitutes integrated analysis of the specific features related to objects on the Earth's surface, namely, shape, size, tone, texture, pattern, shadow, association, and site.

Shape defines the outline of the objects. Objects with certain shapes can be identified from remote sensing image data. Usually man-made objects have geometric and regular shapes, like buildings, roads, agricultural fields, open pits, while natural objects belong to irregular shapes such as forest boundaries, bare lands, etc. (Figure 3.29).

Size gives information about dimensions of the objects. The size of an object can be assessed either relatively by comparing it with other objects or absolutely by considering the number of pixels occupied by the object or making measurements on

Trapezoidal shaped   Rectangular
agricultural field   shaped building

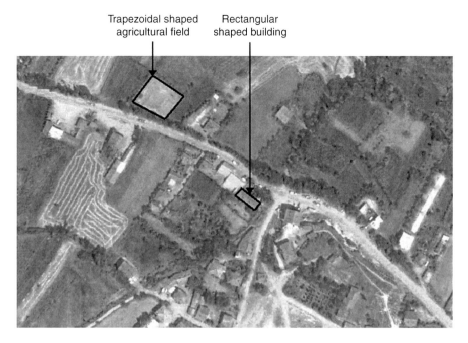

Figure 3.29 Role of shape in visual interpretation

the images. Hence, it is dependent on a scale that is determined by spatial resolution (Figure 3.30).

Tone expresses color of the objects in RGB (true color), NIR-R-G (false color) or IHS color spaces. It is analogous to brightness for a single band (pan) images. The contrast between tones allows the analyst to distinguish and recognize certain objects (Figure 3.31).

Texture is resultant from the tonal variations between the image pixels. High variability between the DN's large tonal variations exhibits a rough texture such as urban area, forest, etc. However, low variability between the DN's (homogeneous tonal values) yields smooth texture such as grasslands, pavements, etc. (Figure 3.32).

The **pattern** reflects information about the spatial configuration of objects. A surface mine layout for example has a pattern with the pit, processing facilities, dump site, and roads, which are different from an urban area with dense buildings and roads (Figure 3.33).

Shadows provide information on shape, size, and height of the objects, illumunation conditions and type of terrain. The size of shadows next to objects gives indications about the relative height of the objects. In addition, when the sun angle is higher (perpendicular or near-perpendicular to objects) shadows are shorter than the ones under lower angles. Moreover, the shadows on the hilly terrain are longer than the ones on the flat terrain (Figure 3.34).

Association refers to the location of certain objects with respect to others. Objects that are difficult to distinguish alone may become identifiable when they are analyzed

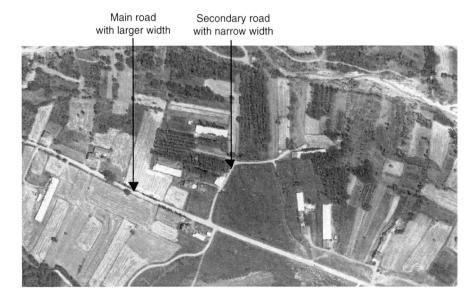

Figure 3.30 Role of size in visual interpretation

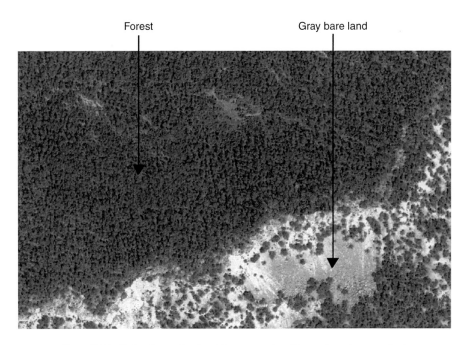

Figure 3.31 Role of tone in visual interpretation (See color plate section)

Low tonal variations of road   High tonal variations of forest

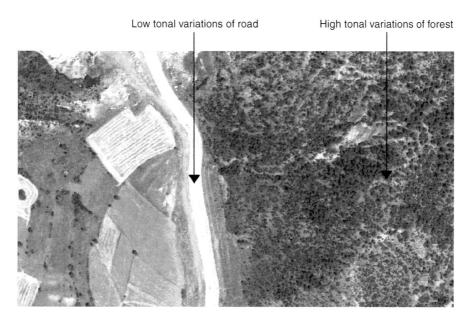

Figure 3.32 Role of texture in visual interpretation

Surface coal mine layout

Figure 3.33 Role of pattern in visual interpretation

with their spatial associations. For example, a surface mine dump site can be identified from bare lands with its connection to the pit (Figure 3.35).

A **site** gives geomorphologic and geographic location information related to objects in the image data. Knowing the image's geographic location helps the analyst make better identification of the objects. For example, certain vegetation types occur in only

Large shadow area due to
tall chimneys of the thermal
power plant

Small shadow area of a
power plant building complex

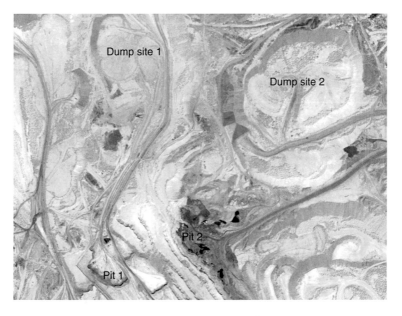

Figure 3.34 Role of shadow in visual interpretation

Dump site 1

Dump site 2

Pit 2

Pit 1

Figure 3.35 Role of association in visual interpretation

certain geographic locations or altitudes, like Mediterranean scrubs which only exist in geographical locations with Mediterranean climates and can grow up to certain altitudes.

Visual image interpretation is highly affected by the spatial resolution. As images with low spatial resolution lack in spatial detail, the role of features mentioned above is not equal due to their limited use in low resolution. For instance, texture and size of the objects may be obscured in low resolution. High spatial resolution images, on the other hand, exploit all of the features, which makes visual interpretation easier than the low resolution image interpretation.

The interpreted image data involves the boundaries, extent, and location of the objects in the form of polygons, lines, and points, which are generally managed through geographic information systems (GIS) either in raster or vector format. Raster data format keeps the data (polygons, lines, and points) in the form of pixels, while vector data format stores them in the form of objects with topological relations. GIS also provide a data management tool for GCPs and other data used in image enhancement and interpretations.

3.3.2 Image interpretation based on digital image analysis

Digital image analysis constitutes interpreting the images based on special algorithms developed for:

* Classification,
* Change analysis,
* Digital elevation models (DEM), and
* Feature extraction.

Classification refers to grouping the object types existing in an image data. It is also called land use and land cover classification, as the objective is to categorize the landscape with its natural and man-made contents. The land cover reflects the naturally occurring object classes on the landscape such as forest, water, sand, etc. The land use, however, involves the human activity on the landscape such as urban, industrial, mine sites. Classification is one of the most widely used remote sensing methods for investigating and monitoring the mining activities and related environments, especially for surface mining operations. The output of a classified image contains pixels with assigned class labels rather than DN's (Figure 3.36).

Regardless of classification methods, the first step in any classification process is to determine the number of classes, their identities and hierarchies, which are essential for an effective classification. After the classification, every pixel of the image should be assigned a class label, which is called a **completion property**. Moreover, there can be some classes, which are the subclasses of a main class like vegetation class and its subclasses of forest and shrubs. Thus, indentifying main and subclasses of them with a defined hierarchy satisfies the **hierarchy property**. The two properties of completion and hierarchy should be satisfied in every classification. Generally, mapping agencies of the countries and international organizations publish hierarchical classes for various mapping scales. The CORINE (Coordination of Information on the Environment) programme of the European Commission for example provides the land use/cover

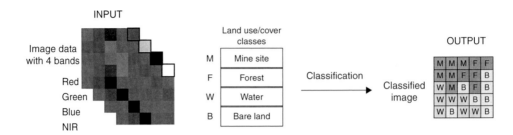

Figure 3.36 The output of image classification (See color plate section)

class hierarchies given in Table 3.2. USGS land use/cover class hierarchies are listed in Table 3.3.

Once the classes are identified with their hierarchies, the next step is to select a classification method. The methods of classification can be viewed in two perspectives. In perspective I, classification is made based on the spatial units namely, pixel-based or object-based classification methods. **Pixel-based classification** assigns a class value to each pixel of the remotely sensed image. **Object-based classification** (also called object-oriented classification) relies on first segmenting the image into homogeneous regions and then assigning the classes to each segment rather than a pixel. Perspective II considers classification methods based on involved supervision like supervised and unsupervised classification methods. **Supervised classification** comprises user interaction during the classification procedure, where a data set called **training data** for each class is selected from the image by the analyst. Then the properties of the training data are extracted to be used in deciding the class type of unclassified image parts. **Unsupervised classification** does not require user interference in the classification procedure. It mainly discovers the relation between the spectral properties inherent in the image and then performs the classification, accordingly. A supervised/unsupervised classification can be carried out either pixel-based or object-oriented. For more complex classification problems hybrid classification algorithms are also adopted, where a mixture of techniques in perspectives I and II are used within the same classification.

The unsupervised classification uses clustering algorithms to group the image data based on spectral values. The only feed back from the user is the number of classes (clusters) to be formed through clustering algorithms. A typical clustering algorithm groups the data based on similarity metrics that are measured usually by calculating the distance between the attributes (i.e. DN's corresponding to each spectral band). The decision of whether a pixel or a region belongs to a certain class (cluster) is made based on evaluating the similarity (distance) of the pixel's DN's in every spectral band to the cluster centre. The pixel's or region's class is decided relatively by comparing the calculated distance with the others. Hence, unsupervised classification is in fact an optimization process, where the optimum class assignment of pixels or regions is searched for. Among the various clustering algorithms, **K-means clustering** is one of the most widely used classified methods in remote sensing image classification. K-means clustering groups the image data, based on k number of land use/land cover classes. Each pixel's or region's class (group) is determined by finding the minimum sum of squared distances between the data (DN) and the associated cluster centroid, which is

Table 3.2 Land use/cover hierarchies of CORINE for remote sensing

Level I	Level II	Level III
1. Artificial surfaces	1.1. Urban fabric	1.1.1. Continuous urban fabric 1.1.2. Discontinuous urban fabric
	1.2. Industrial, commercial and transport units	1.2.1. Industrial or commercial units 1.2.2. Road and rail networks and associated land 1.2.3. Port areas 1.2.4. Airports
	1.3. Mine, dump and construction sites	1.3.1. Mineral extraction sites 1.3.2. Dump sites 1.3.3. Construction sites
	1.4. Artificial non-agricultural vegetated areas	1.4.1. Green urban areas 1.4.2. Sport and leisure facilities
2. Agricultural areas	2.1. Arable land	2.1.1. Non-irrigated arable land 2.1.2. Permanently irrigated land 2.1.3. Rice fields
	2.2. Permanent crops	2.2.1. Vineyards 2.2.2. Fruit trees and berry plantations 2.2.3. Olive groves
	2.3. Pastures	2.3.1. Pastures
	2.4. Heterogeneous agricultural areas	2.4.1. Annual crops associated with permanent crops 2.4.2. Complex cultivation 2.4.3. Land principally occupied by agriculture, with significant areas of natural vegetation 2.4.4. Agro-forestry areas
3. Forests and semi-natural areas	3.1. Forests	3.1.1. Broad-leaved forest 3.1.2. Coniferous forest 3.1.3. Mixed forest
	3.2. Shrub and/or herbaceous vegetation association	3.2.1. Natural grassland 3.2.2. Moors and heathland 3.2.3. Sclerophyllous vegetation 3.2.4. Transitional woodland shrub
	3.3. Open spaces with little or no vegetation	3.3.1. Beaches, dunes, and sand plains 3.3.2. Bare rock 3.3.3. Sparsely vegetated areas 3.3.4. Burnt areas 3.3.5. Glaciers and perpetual snow
4. Wetlands	4.1. Inland wetlands	4.1.1. Inland marshes 4.1.2. Peatbogs
	4.2. Coastal wetlands	4.2.1. Salt marshes 4.2.2. Salines 4.2.3. Intertidal flats
5. Water bodies	5.1. Inland waters	5.1.1. Water courses 5.1.2. Water bodies
	5.2. Marine waters	5.2.1. Coastal lagoons 5.2.2. Estuaries 5.2.3. Sea and ocean

Table 3.3 Land use/cover hierarchies of USGS for remote sensing

Level I	Level II
1. Urban and built-up land	11 Residential 12 Commercial and service 13 Industrial 14 Transportation, communication and utilities 15 Industrial and commercial complexes 16 Mixed urban or built-up land 17 Other urban and built-up land
2. Agricultural land	21 Cropland and pasture 22 Orchards, groves, vineyards, nurseries and ornamental horticultural areas 23 Confined feeding operations 24 Other agricultural land
3. Rangeland	31 Herbaceous rangeland 32 Shrub and brush rangeland 33 Mixed rangeland
4. Forest land	41 Deciduous forest land 42 Evergreen forest land 43 Mixed forest land
5. Water	51 Streams and canals 52 Lakes 53 Reservoirs 54 Bays and estuaries
6. Wetland	61 Forest wetland 62 Nonforested wetland
7. Barren land	71 Dry salt flats 72 Beaches 73 Sandy areas other than beaches 74 Bare exposed rock 75 Strip mines, quarries and gravel pits 76 Transitional areas 77 Mixed barren land
8. Tundra	81 Shrub and brush tundra 82 Herbaceous tundra 83 Bare ground tundra 84 Wet tundra 85 Mixed tundra
9. Perennial snow and ice	91 Perennial snowfields 92 Glaciers

the mean value of each class's data values (DN). Each time a pixel or region assigned to a cluster, cluster centroids are re-evaluated. The main steps of K-means clustering algorithms are as follows:

1   Determine k, which is the number of classes.
2   Calculate an initial cluster centroid by obtaining a sample set which has k number of clusters with a single element (the data value of each element will be the initial centroid). It can be obtained by selecting arbitrary pixels or regions. Then assign

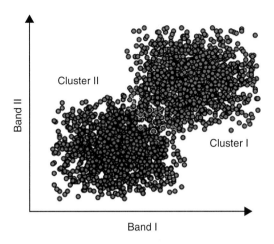

Figure 3.37 K-means clustering for two classes (k = 2) (See color plate section)

each of the remaining samples set to the nearest centroid. Re-compute the centroid after each assignment. By this way, initial sets of k clusters are obtained.

3   Select each pixel or region and compute its distance to each cluster centroid. Assign the pixel/object to the cluster with minimum distance and re-evaluate the cluster centroid after assignment.

4   Repeat step 3 until there is no unclassified pixel/region is left.

Figure 3.37 shows k-means classification for two bands of a remotely sensed image with two classes (k = 2). As most of the remote sensing images contain multiple bands, the clusters obtained for such images are n-dimensional objects.

The supervised classification, on the other hand, contains a learning process through training data. For an efficient and accurate classification, the training data related to the classes should be selected from the same image. The main reason for this is that training data obtained from other images exhibit different properties resulting in poor learning, as the atmospheric, illumination, and sensor conditions may not be the same as the image to be classified. In the supervised classification the basic statistical properties of each class's spectral values are determined from the training data. Then each pixel's/region's class is determined by obtaining a similarity measure for the individual pixel's/class's spectral signatures and the training set statistics. Finally, according to the degree of similarity, each pixel/region is assigned a class label. For this reason, the training set should satisfy the following properties for an accurate classification.

1   **Completeness:** Training data should contain all spectral bands to be used in classification. As many objects on the Earth have various spectral reflectance characteristics for different regions of the electromagnetic spectrum, for every land use/cover class, all the statistical measures representing spectral properties of every band of the image should be included in the training data.

2 **Representativeness:** Training data should be representative of all the available spectral variability for a given class. For example, in an open pit mine area, the dump site and the pit may have different spectral reflectance characteristics due to differences in texture and exposed rock features. If a classification process involves a class of mine site, the training set should involve pixels/regions both from the dump site and the pit.

3 **Sufficiency:** Theoretically, the number of training data to be used in a supervised classification based on probabilistic approaches is $n + 1$, where n is the number of image bands (Lillesand and Kiefer, 2000). Hence if a 4-band multispectral data is used for classification, there should be at least five training data. In practice, the required number of training data ranges between 10n to 100n (Lillesand and Kiefer, 2000). There are other suggestions related to the number of training data. For example, the pixels of at least 30n/each class are suggested by Mather (2004). For a detailed discussion of the required number of training data, see Tso and Mather (2009). In general, the higher the number of training set, the better is the classification results.

4 **Spatial distribution:** The training data from every classification class should be collected from every possible part of the image. Collecting a set of a large number of pixels/regions for a single class from only one part of the image should be avoided.

The training data, which satisfy the four criteria, are mainly obtained by the analyst. The previously collected field data, GIS layers, existing maps of the site are the other sources of the training data. In any case, the training data collection is time consuming and requires some knowledge of the area from which the image is taken.

The most frequently used supervised classification algorithms in remote sensing applications can be grouped into five major classes:

1 Probabilistic classification algorithms,
2 Machine-learning-based classification algorithms,
3 Artificial intelligence-based classification algorithms,
4 Decision tree classification algorithms, and
5 Fuzzy set-based classification algorithms.

The probabilistic classification algorithms mainly rely on extracting statistical information from the training data and finding a statistical measure as the indication of each pixel's/region's similarity to every class. Then the class label having the similarity value is assigned to the pixel/region. There are mainly three algorithms in probabilistic classification namely:

• Minimum distance to means classification,
• Parallelepiped classification, and
• Maximum likelihood classification.

In minimum distance to means classification, the similarity measure is the mean of the spectral values for each class in the training data. Hence, the comparison is performed between each pixel's/region's spectral values and the mean values of each class. The pixel/region having the closest spectral values to the mean of the classes is assigned

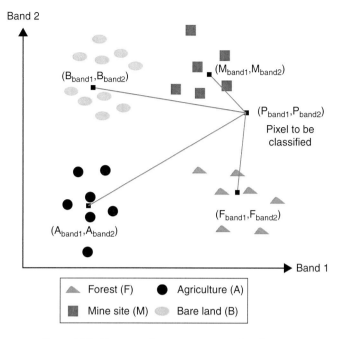

Figure 3.38 Minimum distance to means classification

the class label. The similarity between the class mean of the pixel/region is determined by finding the minimum differences between mean of the spectral values in each class for each spectral band and the pixel's/region's value of each band. For a two-band case, classification reduces to finding the minimum distance from the pixel's/region's location to the location of each class mean. Figure 3.38 shows the concept in two dimensional cases with classes of forest (F), mine site (M), agriculture (A), and bare land (B). If the pixel (P) has coordinates of (P_{band1}, P_{band2}) and the coordinates of F, M, A, and B are (F_{band1}, F_{band2}), (M_{band1}, M_{band2}), (A_{band1}, A_{band2}), and (B_{band1}, B_{band2}), respectively, then the distances from P to F (d_{PF}), P to M (d_{PM}), P to A (d_{PA}), and P to B (d_{PB}), are evaluated using Eq. 3.11. The minimum distance of d_{PF}, d_{PM}, d_{PA}, d_{PB} gives the class label of P. In multispectral image data the mean of each class forms an n-dimensional vector, where n is the number of spectral bands in the image data. Hence the vector of spectral values for any pixel/region is compared with the vector of each class's mean. The main disadvantage of this classification is the fact that equal or very close distances between the classes may result in a high degree of misclassification.

$$
\begin{aligned}
d_{PF} &= \sqrt{(P_{band1} - F_{band1})^2 + (P_{band2} - F_{band2})^2} \\
d_{PM} &= \sqrt{(P_{band1} - M_{band1})^2 + (P_{band2} - M_{band2})^2} \\
d_{PA} &= \sqrt{(P_{band1} - A_{band1})^2 + (P_{band2} - A_{band2})^2} \\
d_{PB} &= \sqrt{(P_{band1} - B_{band1})^2 + (P_{band2} - B_{band2})^2}
\end{aligned}
\tag{3.11}
$$

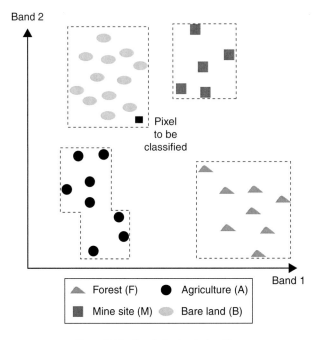

Figure 3.39 Parallelepiped classifier

The similarity measure for parallelepiped classifier is the range (maximum–minimum) of spectral values for each class in the training data. In the two-dimensional case with band1 and band2, the range forms a region of a rectangle, while it is a rectangular prism in the three-dimensional case. For an n-dimensional case the region of each class is called parallelepiped. The class of a pixel/region is assigned if its spectral values fall in a specific class (Figure 3.39). When class boundaries have a high degree of overlapping regions, deciding on the class label of a pixel/region becomes problematic leaving the pixel/region unclassified in this classification algorithm. Such unclassified pixels/regions can either be classified by the user or require an additional run of another algorithm such as minimum distance to means classification.

The maximum likelihood classification adopts the probability values as the similarity measure, where the shortcoming of minimum distance to means and parallelepiped classification algorithms are overcome. The probability distributions of each class for spectral bands are obtained from the training data. Then the probability of a pixel's/region's spectral values belonging to each class is computed. The class label of the pixel/region is assigned according to the highest probability value. In the two-dimensional case, probability values for each class form equal probability contours (Figure 3.40). If the training data is sufficient to represent all the variability in the spectral values of each class, maximum likelihood classification is one of the most accurate classification methods. It assigns a class label to every pixel/class in the image. However, the maximum likelihood classification can computationally be complex for a large number of classes and spectral bands.

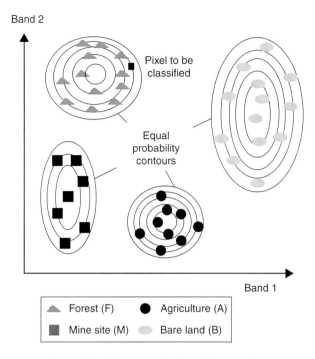

Figure 3.40 Maximum likelihood classification

Support Vector Machine (SVM) is among the most widely used **machine-learning-based classification algorithms**. SVM forms a hyperplane which has the maximum separation margin between the two classes (Figure 3.41). As can be seen from Figure 3.41 there are an infinitely large number of hyperplanes, with some examples of HP1, HP2, and HP3. However, HP2 has the widest margin of separation. A subset of training data, which is called support vectors, defines the hyperplane with maximum margin. It has been showed by Huang *et al.* (2002) and Pal and Mather (2005) that SVM performs better in remote sensing applications than maximum likelihood classification.

Artificial Neural Networks (ANN) forms the backbone of the artificial intelligence-based classification. The ANN determines the hidden relations and rules from the training data, which is also called learning. Once the network learns from the training data, it uses the extracted rules and relationships to assign class labels to pixels/regions. There are basically five types of ANN namely, multilayer perception with back-error propagation, counter propagation, self-organizing maps (SOM), Hopefield, and adaptive resonance theory-based ANN (ARTMAP).

Fuzzy set-based classification algorithms provide better treatment of uncertainties in the classification, which results in mixed classes or hardly separable classes. In this type of classification being a member of a class is defined by fuzzy sets, where the set properties are obtained from the training data. As is also stated by Tso and Mather (2009), use of Fuzzy-set based classification algorithms have become more

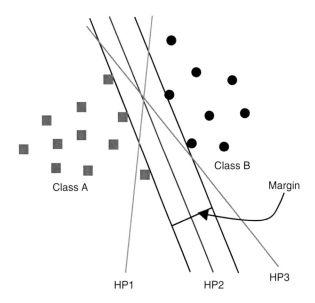

Figure 3.41 Hyperplanes separating the classes A and B

popular due to their ability to handle mixed class problems. Fuzzy C-means, fuzzy maximum likelihood, and fuzzy rule-based classification algorithms are the typical types of fuzzy set-based classification algorithms.

Decision tree classification algorithms are based on a hierarchical division of image data for defined class hierarchies, which can be constructed manually by the user or automatically using statistical properties of the spectral bands. A decision tree is composed of three main components namely, root, interior, and terminal (leaf) nodes as well as decision rules for identification of tree branches to be selected. The class assignment procedure starts from the root node and reaches to a terminal node through selecting tree branches with interior nodes using decision rules.

A more detailed description of supervised and unsupervised classification algorithms can be found in Tso and Mather (2009). Currently, classification of remote sensing data has mainly been performed using hybrid approaches, where pixel-based and object-based methods have been integrated with the above mentioned classification algorithms. Typical examples are Shackelford and Davis (2003), Wang and Qin (2005), Weih and Riggan (2009), Buddhiraju and Rizvi (2010), and Yang and Ma (2010).

Once the classification is carried out it should be coupled with the assessment of the **classification accuracy**, which provides information about the reliability of the final classification output. In order to perform accuracy assessment, a set of data with known (true) classes, also called **ground truth**, is required. Ground truth data can be obtained from field studies, existing maps or higher resolution images with detectable classes. Then the classification accuracy is assessed by evaluating the match between the ground truth data and correspondences classification outcomes. This

Table 3.4 Typical error matrix

Classification Output	Ground Truth Data				
	F (Forest)	M (Mine Site)	A (Agriculture)	B (Bare Land)	Row total
F	356	0	13	0	369
M	3	73	27	68	171
A	19	28	152	0	199
B	0	12	2	98	112
Column total	378	113	194	166	851

User's accuracy		Producer's accuracy	
F	356/369 = 0.96	F	356/378 = 0.94
M	73/171 = 0.43	M	73/113 = 0.65
A	152/199 = 0.76	A	152/194 = 0.78
B	98/112 = 0.88	B	98/166 = 0.59

Overall accuracy = (356 + 73 + 152 + 98)/851 = 0.80

comparison is systematically done with the help of the so-called **error matrix** (confusion matrix/contingency table). A typical error matrix is given in Table 3.4 for the classes of forest (F), mine site (M), agriculture (A), and bare land (B). The error matrix involves the correct and incorrect class assignments by the classification algorithm for the ground truth data. The diagonal elements of the error matrix provide the number of pixels/regions classified correctly. The off-diagonal elements present the misclassified ones. The non-diagonal column elements are called omission errors while the non-diagonal row elements are called commission errors. Omission errors are the pixels/ regions, which should have been classified but are omitted from the considered class (Table 3.4). For example although the ground truth data has a class of mine site, it is not labeled by the classification as mine site class. The commission errors, on the other hand, are the pixels/regions, which are assigned a class label by the classifier but not have the same class label in the ground truth data (Table 3.4). For example, the pixel/region classified by the classifier as mine site does not actually belong to the class of mine site in the ground truth data. The ratio of correctly classified pixels/regions for each class in the columns and rows of the error matrix is called producer's and user's accuracy, respectively. Error matrix is also used for obtaining statistical measures of accuracy. The most widely used statistic is the **Kappa statistic** (Eq. 3.12).

$$\kappa = \frac{P(A) - P(E)}{1 - P(E)} \qquad (3.12)$$

In Eq. 3.12, the expression $1 - P(E)$ reflects the amount of classification agreement with the ground truth that can be achieved better than chance. $P(A) - P(E)$, on the other hand, reflects the actual agreement between the classification and ground truth. For N number of ground truths and k number of classes with n × n error matrix, each element of the error matrix (n_{ij}) has indices $i = 1, \ldots, n$ and $j = 1, \ldots, k$. Then $P(A)$ and $P(E)$ can be computed from Eq.'s 3.13 and 3.14.

$$P(A) = \frac{1}{Nn(n-1)} \left(\sum_{i=1}^{n} \sum_{j=1}^{k} n_{ij}^{2} - Nn \right) \tag{3.13}$$

$$P(E) = \left[\frac{\left(\frac{1}{n} \sum_{j=1}^{k} n_{ij} \right)}{Nn} \left(\sum_{i=1}^{N} n_{ij} \right) \right]^{2} \tag{3.14}$$

In a more practical manner, the numerator of the Kappa statistic can be calculated by multiplying the total number of elements in the error matrix (N) with the sum of diagonal elements of the error matrix and then subtracting the sum of multiplication of row and column sums. The denominator of the Kappa statistic is N^2 subtracted by the sum of multiplication of row and column sums. For the error matrix given in Table 3.4, the Kappa statistic is computed as:

$N = 851$
Sum of diagonal $= 356 + 73 + 152 + 98 = 679$
Sum of row and column sum multiplications
$\quad = (369 * 378) + (171 * 113) + (199 * 194) + (122 * 166)$
$\quad = 138442 + 19323 + 38606 + 20252 = 216623$
$N^2 = 724201$
$\kappa = \dfrac{851 \cdot 679 - 216623}{724201 - 216623} = \dfrac{361206}{507578} = 0.71$

Change analysis involves comparing at least two different remotely sensed images obtained in different time periods. The algorithms that can be used for change analysis are mainly of two types. The first group of algorithms mainly compares two different raw images using arithmetic operations like taking the difference or ratio of the spectral bands of the two or more images. The values close to zero and one in the resulting change image indicate unchanged pixels, for subtraction and ratioing, respectively. The second group of change detection algorithms compares processed images for the two or more different time periods like comparing vegetation indices, classification results, and principal component analysis. Prior to any change analysis application, the following pre-image processing analyses should be performed in order to have a better change analysis accuracy.

1   Precise geometric rectification analysis is required for accurate spatial pixel matching between the images. Inadequate geometric rectification yields unchanged areas as if they are changed since the spectral values corresponding to the same area will not be the same for the same pixel.
2   Atmospheric correction for all the images is necessary in order to eliminate atmospheric effects in the spectral values.

3   If the images for the two or more different dates do not have the same spatial resolution and/or spectral band combinations, their spatial and spectral resolution should be matched by image enhancement and resampling methods.

The result of change analysis is usually given in the form of a change map and a change matrix. An overview of change analysis methods can be found from Singh (1989). In addition, Schimidt and Glaeser (1998) give typical application of change analysis for open cast coal mines.

The digital elevation model (DEM) of an area from remote sensing can either be obtained from stereo pairs of the same area with at least 60% overlapping areas or from aerial/satellite images or LIDAR data or from radar interferometry. DEM generation is mainly obtaining the terrain height for certain points on the image and interpolating the height values to create a continuous height surface. Acquiring height information from stereo images relies on the fundamental concept of image **parallax**. Parallax is the change of fixed (stationary or not moving) objects in the image by viewing them in different positions. The object viewed in different directions exhibits different locations in each stereo image. Then the parallax of a fixed point A, (P_a) is defined by the difference between the x coordinate of the A in the left image (x_a) and the x coordinate of the right image (x'_a) (Eq. 3.15).

$$P_a = x_a - x'_a \qquad (3.15)$$

For the distance of B, which is defined by the horizontal distance between the center of the right and left images (also called the image base), camera lens of f, height of H at which the images are taken, the height (elevation) of point A (h_a) is given in Eq. 3.16.

$$h_a = H - \frac{Bf}{P_a} \qquad (3.16)$$

If an image is taken with a 200 mm lens from 1500 m height with 700 m base, using the measured x coordinates of a point A on the two images, (x_a) and (x'_a), which is 35.24 mm and −67.73 mm, respectively, the elevation of the point A (h_a) is:

$$h_a = 1500 - \frac{35.24 \cdot 200}{[35.24 - (-67.73)]} = 1431 \, m$$

Once a set of height values is found from the image, the elevation of every image pixel can be found by interpolation. Finally, a DEM forms a gray scale image, where the pixel values are elevation values, not the spectral values. A typical DEM obtained by processing of the stereo aerial photos for a mine site is given in Figure 3.42.

LIDAR stands for **L**ight **D**etection **a**nd **R**anging, which is based on LASER (**L**ight **A**mplification by **S**timulated **E**mission of **R**adiation) technology coupled with Global Positioning Systems (GPS) and **I**nertial **M**easurement **U**nit (IMU) of the aircraft's Inertial Navigation System (INS) for attitude measurement. When the LIDAR system is ground-based, in which the laser beam is sent to the target from the ground, not from the aircraft or helicopter, IMU is unnecessary. Ground-based LIDAR systems are mostly used for relatively small regions like mine sites. For large coverage of areas, airborne LIDAR systems are preferred. The pulses of laser light are sent

Figure 3.42 DEM image

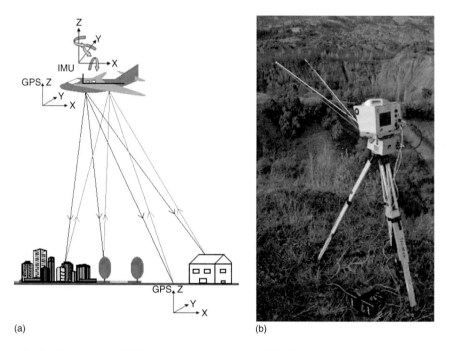

(a)　　　　　　　　　　　　　　　　　　(b)

Figure 3.43　(a) Airborne LIDAR system, (b) ground-based LIDAR System (See color plate section)

from a LIDAR system to the target on the ground and then the time between sending and reflectance incidents of the laser beam is measured. As the speed of the light is constant, the distance to the target is calculated from the record of elapsed time for sending the laser beam and collecting its reflection. Using additional information of the laser sensor's viewing angle, position height, and orientation obtained from GPS and IMU, a point's location and elevation can be measured precisely (Figure 3.43).

Figure 3.44 Typical point cloud data from LIDAR

Hence, positional and elevation data for a specified density of points are collected with a vertical accuracy ranging between 10–50 cm by sending successive pulses of laser. The pulse rates and point density range between 75,000–100,000 pulses/second and 1–10 points/m respectively, depending on the required level of detail. Although LIDAR systems collect point clouds (Figure 3.44) with quite precise position and elevation information, the processing data and obtaining DEM from them is relatively costly due to the requirement of efficient algorithms for large datasets as well as noise elimination processes. Once the LIDAR data in the form of point clouds is processed, DEM surface is produced by interpolation.

Radar interferometry has similar principles to LIDAR, where instead of laser light, electromagnetic energy in the microwave region of the electromagnetic spectrum (Figure 2.4) is used. A Synthetic Aperture Radar (SAR) system records the return time and strength of the reflected radar wave as well as the point on the wave signal when the returned wave is received. The point of the sinusoidal wave at which the returned wave is received is called **phase**. The comparison of phase difference recorded on the reflected radar waves obtained from the same location on the Earth provides the elevation of a point of the Earth. The elevation values are mainly evaluated by acquiring phase differences and accurately registering them. Registration is basically performed by overlaying the phase values and then subtracting them from each other. The image after this process is called **interferogram**, which forms the basis of elevation calculations. For a detailed description the reader may refer to Small *et al.* (1998) and Wegmüller *et al.* (2009). The detailed description for methods of DEM extraction from radar images is given in Chapter 4.

Feature extraction is mainly extracting boundaries of specific features like building, road, railroad, airport, etc. from the images. These algorithms are specifically developed for extracting each feature such as building extraction algorithms, road extraction algorithms, etc. and the features are mainly man-made objects. For this reason such types of image interpretation are quite rare in mining applications. Hence, they are not in the scope of this book.

3.4   SOFTWARE FOR PROCESSING OF REMOTELY SENSED DATA

Analysis of the remote sensing images for various applications can be eased using specifically designed software products for this purpose. These software products would be commercially available ones in the market or freely available open-source ones that have their codes open to the remote sensing community. Among the various products, the most commonly used commercial software products are ERDAS-LPS, PCI-Geomatica, ENVI-IDL, Z/I, GAMMA, SARscape, OSSIM (http://www.ossim.org/ OSSIM/OSSIMHome.html), ORFEO (http://www.orfeo-toolbox.org/otblive/), Multi-Spec (http://cobweb.ecn.purdue.edu/~biehl/MultiSpec/), IDIOT (http://srv-43-200.bv. tu-berlin.de/idiot), DORIS (http://doris.tudelft.nl/). ERDAS-LPS, PCI-Geomatica and ENVI-IDL, Z/I provide capabilities for processing optical, radar and lidar data with associated modules. GAMMA and SARscape are specifically designed for processing radar (microwave) images. OSSIM and ORFEO are mainly for optical remote sensing open source products, while DORIS and IDIOT are open-source software for processing of radar images.

In addition to remote sensing software products, remote sensing analysis is supported by Geographic Information Systems (GIS) software. GIS are designed for sorting, manipulating, analyzing, and visualizing the spatial data. For this reason, the remote sensing analysis products feed GIS data layers. Moreover, the management and preparation of ground truth data can be handled better in GIS if the existing maps are used. Nowadays, due to the widespread use of remote sensing image products, many GIS software products contain various image analysis functions. Besides, the remote sensing analysis products can be manipulated better in GIS software so as to publish them with efficient visualizations.

Google Earth can be considered another complementary product for remote sensing analysis, where the existing data in Google Earth may serve for ground truth and GCP data to be used as input to various image analysis algorithms.

As all the remote sensing image data have geographic coordinates, field data collection to be used in accuracy assessment or geometrical image correction requires ground truth or GCP data to be obtained with geographical coordinates. For this purpose, GPS devices with their software play an important role in the set of remote sensing supporting technologies.

REFERENCES

Aranof, S. (2005) *Remote Sensing for GIS Managers*. California, USA, ESRI Press.

Baret, F. & Guyot, G. (1991) Potentials and limits of vegetation indices for LAI and APAR assessment. *Remote Sensing of Environment*, Vol. 35, 161–173.

Buddhiraju, K.M. & Rizvi, I.A. (2010) Comparison of CBF, ANN and SVM classifiers for object based classification of high resolution satellite images. *Proceedings of Geoscience and Remote Sensing Symposium (IGARSS)*. pp. 40–43.

Chen, J. M. (1996) Evaluation of vegetation indices and a modified simple ratio for boreal applications. *Canadian Journal of Remote Sensing*, Vol. 22, 229–242.

Clevers, J. G. P. W. (1988) The derivation of a simplified reflectance model for the estimation of leaf area index. *Remote Sensing of Environment*, Vol. 35, 53–70.

Clevers, J. G. P. W. (1991) Application of the WDVI in estimating LAI at the generative stage of barley. *ISPRS Journal of Photogrammetry and Remote Sensing*, Vol. 46, 37–47.

Fassnacht, K. S., Gower, S. T., MacKenzie, M. D., Nordheim, E. V. & Lillesand, T. M. (1997) Estimating the leaf area index of north central Wisconsin forests using Landsat Thematic Mapper. *Remote Sensing of Environment*, Vol. 61, 229–245.

Glenn, E.P, Huete, A.R., Nagler, P.L. & Nelson, S.G. (2008) Relationship between remotely-sensed vegetation indices, canopy attributes and plant physiological processes: What vegetation indices can and cannot tell us about the landscape, Sensors, Vol. 8, 2136–2160.

Goel, N. S. & Qi, W. (1994) Influences of canopy architecture on relationships between various vegetation indices and LAI and FPAR: A computer simulation. *Remote Sensing of Environment*, Vol. 10, 309–347.

Gong, P., Pu, R., Bigging, G. S. & Larrieu, M. R. (2003) Estimation of forest leaf area index using vegetation indices derived from hyperian hyperspectral data, *IEEE Transactions on Geosciences and Remote Sensing*, Vol. 41, 1355–1362.

Gonzalez, R.C. & Woods, R.E. (1992) *Digital Image Processing*. Reading, MA, Addison-Wesley.

Haralick, R.M., Shanmugan, K. & Dinstein, I. (1973) Textural features for image classification. *IEEE Transactions on Systems, Man and Cybernetics*, SMC-3, 610–621.

Huang, C.L, Davis, S. & Townshed, J.R.G. (2002) An assessment of support vector machines for land cover classification. *International Journal of Remote Sensing*, 23, 725–749.

Huete, A. R. (1988) A soil adjusted vegetation index (SAVI). *Remote Sensing of Environment*, Vol. 25, 295–309.

Jain, A.K. (1989) *Fundamentals of Digital Image Processing*. Englewood Cliffs, NJ, USA, Prentice Hall.

van Leeuwen, W. J. D. & Huete, A. R. (1996) Effects of standing litter on the biophysical interpretation of plant canopies with spectral indices. *Remote Sensing of Environment*, 55, 123–138.

Lillesand, T.M. & Kiefer, R.W. (2000) *Remote Sensing and Image Interpretation*. USA, John Wiley and Sons. Inc.

Mather, P.M. (2004) *Computer Processing of Remotely Sensed Images: An Introduction Data*, Third Edition. Chichester, John Wiley and Sons.

Pal, M. & Mather, P.M. (2005) Support vector machines for classification in remote sensing. *International Journal of Remote Sensing*, 26, 1007–1011.

Pratt, W.K. (1991) *Digital Image Processing*, Second Edition. New York, NY, USA, John Wiley and Sons.

Roujean, J. L. & Breon, E. M. (1995) Estimating PAR absorbed by vegetation from bidirectional and reflectance measurements. *Remote Sensing of Environment*, Vol. 51, 375–384.

Sabins, Jr., F. F. (1987) *Remote Sensing; Principles and Interpretations*. New York: W. H. Freeman.

Schmidt, H. & Glaesser, C. (1998) Multitemporal analysis of satellite data and their use in the monitoring of the environmental impacts of open cast lignite mining areas in East Germany. *International Journal of Remote Sensing*, 19, 2245–2260.

Schowengerdt, R.A. (1997) *Remote Sensing Models and Methods for Image Processing*, Second Edition. San Diego, USA, Academic Press.

Shackelford, A.K. & Davis, C.H. (2003) A combined segmentation and pixel based classification approach of Quickbird imagery for land cover mapping. *IEEE Transactions on Geoscience and Remote Sensing*, 41 (10).

Singh, A. (1989) Digital change detection techniques using remotely sensed data. *International Journal of Remote Sensing*, 6, 989–1003.

Small, D., Pasquali, P., Holecz, F., Meier, E. & Nüesch D. (1998) Experiences with multiresolution and multifrequency InSAR Height Model Generation. *Proceedings of IEEE-IGARSS'98*. pp. 2671–2673.

Tarabalka, Y., Chanussot, J. & Benediktsson, J. (2010) Segmentation and classification of hyperspectral images using watershed transformation. *Pattern Recognition*, 43 (7), 2367–2379.

Tso, B. & Mather, P.M. (2009) *Classification Methods for Remotely Sensed Data*, Second Edition. Boca Raton, Taylor and Francis Group.

Tucker, C. J. (1979) Red and photographic infrared linear combinations for monitoring vegetation. *Remote Sensing of Environment*, Vol. 8, 127–150.

Wang, J., Li, D. & Qin, W. (2005) A combined segmentation and pixel based classification approach of Quickbird imagery for land cover mapping. *Proceedings of SPIE*, vol. 6044, 60440U.

Wegmüller, U., Santoro, M., Werner, C., Strozzi, T., Wiesmann, A. & Lengert, W. (2009) DEM generation using ERS–ENVISAT interferometry. *Journal of Applied Geophysics*, 69, 51–58.

Weih, R.C. & Riggan, N.D. A Comparison of Pixel-based versus Object-based Land Use/ Land Cover Classification Methodologies, http://www.featureanalyst.com/feature_analyst/ publications/success/comparison.pdf.

Yang, H., Du, Q. & Ma, B. (2010) Decision fusion on supervised and unsupervised classifiers for hyperspectral imagery. *Geoscience and Remote Sensing Letters, IEEE*, 7 (4), 875–879.

Chapter 4

Remote sensing in subsidence monitoring

This chapter covers the remote sensing techniques utilized in mine subsidence monitoring. It provides the definition, basic mechanisms, and also causes and results of mine subsidence. Then important techniques that have been utilized through time in mine subsidence monitoring are briefly described. Among these mine subsidence monitoring alternatives, remote sensing techniques as passive (optical) and active (microwave) remote sensing are focused throughout the chapter. Different approaches, such as uses of satellite images, aerial photographs, InSAR, and DInSAR are explained and the methodologies are described briefly. Finally, a case study is presented to show the utilization of aerial photographs in mine subsidence monitoring for measuring and characterizing the surface movement changes in underground coal mining environment.

4.1 AN OVERVIEW

Subsidence is one of the most noticeable environmental impacts associated with mining. Mine subsidence is a movement of the surface in the form of small scale collapses, such as, sinkholes or troughs on the surface and regional settlements as a result of settlement of the overburden due to disintegration or failure of underground mine workings (Figure 4.1). Subsidence can have severe economical, technical, social, and

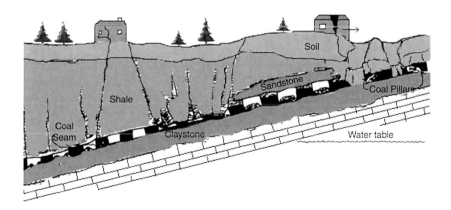

Figure 4.1 Modes of subsidence

environmental impacts. The most common impacts of mine subsidence are observed on surface structures, bridges, buildings, services and communications lines, and agricultural land through the disruption of drainage patterns and variations of gradient. The extent of potential impacts of subsidence depends on the type of mining method, geology of the deposit and the overburden, attitude of the mineral deposits, and mitigation measures. Generating appropriate mitigation measures necessitates continuous monitoring and investigation of the mine area and its surroundings. In this sense, remote sensing can provide an effective tool in measuring and interpreting subsidence quickly, accurately, and simply.

4.2 MECHANISM OF MINE SUBSIDENCE

At depth rock strata are considered to be in an equilibrium state with hydrostatic state of stress. The natural stress field is altered when underground mining or other processes create a void within the formation. The alteration in stress state results in the deformation and displacement of the rock strata throughout a zone which can extend to the surface depending on the extension of the size and depth of the excavation (Wright and Hoffman, 1995). Thus, the cavity artificially created underground by the extraction of ore removes the natural support from the overlying strata. As a result, successive layers of rock over the mine workings bend under the influence of gravity, finally the movement reaches the upper Earth surface (Kratzsch, 1983). When an excavation is extensive, the immediate roof strata break and cave into void. The broken material expands in volume after it caves. It then provides some degree of support to the overlying strata. The caving continues to propagate upward until it reaches a competent stratum that is capable of bridging the void or until caving ceases due to the development of an arched caving zone in which the differences in stress state have been relieved. The mechanism of mine subsidence is illustrated in Figure 4.2 and three main phases of the mine subsidence are summarized in Figure 4.3.

Kratzsch (1983) claimed that mine subsidence is essentially associated with six zones of movement in the rock mass independently of time. They are (Kratzsch, 1983):

a) The floor layers, which are elastically upwards on the relief of the perpendicular load (Figure 4.2);
b) The seam and waste or packing layer, which is inelastically compressed both by the front and back abutment pressures (Figure 4.2);
c) The immediate roof layer, which detaches itself from the more rigid main roof (Figure 4.2);
d) The main roof settles gradually (Figure 4.2);
e) The intermediate zone, consisting of thick beds of solid rock, which sag mostly elastically and detached horizontally along bedding planes (Figure 4.2);
f) The surface zone of loose overburden layers, which behaves plastically (Figure 4.2).

The state of stress in rock strata is given in Figure 4.4. As Figure 4.4 illustrates the maximum subsidence occurs at the center of the subsidence profile. The foundation in the compression zone will be under the excessive compressive forces and that in the horizontal extension zone will be under excessive tensile forces.

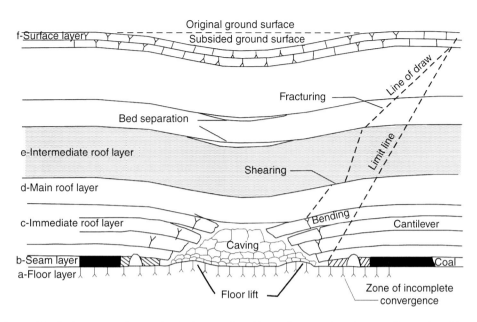

Figure 4.2   Cross-sectional view of subsidence induced by underground coal mining (Reproduced from Shadbolt, 1977)

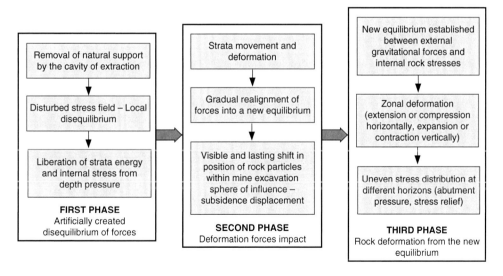

Figure 4.3   Main phases of subsidence

4.3   MAIN CAUSES OF SUBSIDENCE

Subsidence mechanism initiates with the settlement of voids left from extraction of mineral resources using underground mining methods, and ends with the propagation of this settlement from underground layers to the surface. The main factors causing this

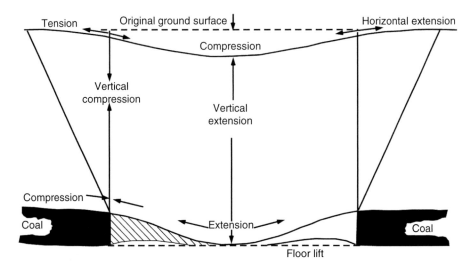

Figure 4.4 The state of stress in strata overlying an area of extraction (Reproduced from Shadbolt, 1977)

movement are classified into two groups: (i) factors causing the initiation of the failure of void spaces along the ore bed and (ii) factors related to the structure and behavior of ore and overburden layers. These factors, which have an impact on the generation and speed of the subsidence mechanism, are grouped as constant and variable factors.

Constant factors

Constant factors are inherent to the geology of the subsidence region including mineralogy and petrography of the hanging wall and footwall, stratigraphy of the waste rock layer, geomechanical and hydrological conditions of the ground, surface topography and vegetation, thickness and dip of the ore bed, and completed production.

The physical properties of the overburden strata, as one of the most important factors, control the shape and size of the subsidence. Sandstones and limestones behave in a more brittle manner and are prone to bridging or sudden failure under loading (Wright and Hoffman, 1995).

Variable factors

Variable factors are factors related to the underground production face, operational factors related to mining methods utilized, such as room and pillar, caving, longwall mining methods, and production rate.

Subsidence, if it cannot be controlled, could result in various problems in the subsidence region. They are:

i)   Damage to buildings, roads, water distribution systems, sewage, energy and communication facilities, which are the physical consequences of subsidence induced by horizontal and vertical movement and replacement of the ground.

ii) Disruption of the underground and surface water drainage, erosion, rock falls, and landslides.

iii) Increase in the frequency and extent of above mentioned problems if they are already present.

iv) Difficult surveying and mapping conditions yielding complexities in land use and ownership with various economical and social problems.

4.4 MINE SUBSIDENCE DAMAGE

Effective assessment of mine subsidence is usually required to consider the impact of particular workings on a given structure. It was claimed that the individual elements of ground movement have different effects and varying importance for different types of structure (Bell, 1975). According to Bell (1975) mine subsidence is the most important type of movement in low-lying areas which are subject to flooding and drainage problems. Even a few centimeters of subsidence can sometimes result in very serious damage. Tilt is also a concern knowing the fact that it may damage drainage works and affect highways, canals, and rail tracks, chimneys, and it brings about the disruption of industrial machinery.

The impact of vertical movement changes the original topography of the surface and causes the disruption of structures at the surface. This disruption may result in bending and shear stresses on the structures depending on the material characteristics, the type of the structures and the geometry of the structures. When these stresses exceed elastic limits, then tension cracks and shear cracks may occur in critical locations. The impact of horizontal movement causes ground surface changes its position relative to the foundation. The amount of deformation on the foundation depends on the frictional force which arises from the interaction between the foundation and the ground due to the shape and depth of the foundation.

Damage to buildings and conventional structures is generally caused by differential horizontal movements and the concavity and convexity of the subsided profile resulting in compression and extension in the structure itself (Figure 4.5). Convex and concave profiles result in bell-shaped and V-shaped fractures, respectively. Many materials and structures are stronger in compression than tension. Therefore, most buildings require approximately twice the amount of compression to develop damage comparable to that caused by a given amount of tension. Typical mining damage starts to appear in conventional structures when they are subjected to effective strains of 0.5–1.0 (Bell, 1975).

A scale of damage resulting from mine subsidence is listed in Table 4.1.

The impacts of mine subsidence and the resultant damage are controlled by the following precautionary measures.

1 Precautionary measures built into new structures in mining areas,
2 Preventive works applied to existing structures that are to be affected by underground mining,
3 Mining design incorporating special underground layouts, and
4 Any combination of 1–3.

The abovementioned influencing factors and detrimental impacts of subsidence implies that subsidence monitoring is essential for legislative requirements, subsidence prediction, maximized coal extraction, structural design, risk management, and environmental monitoring (Ge *et al.*, 2007).

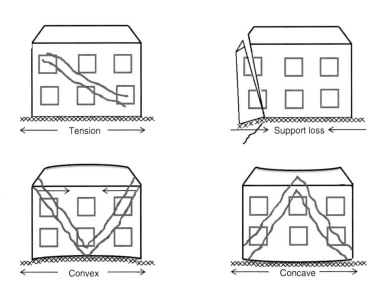

Figure 4.5 Damage types on structures

Table 4.1 National Coal Board classification of subsidence damage (Belly, 1975; Wright and Hoffman, 1995)

Change in length of structure (mm)	Class of damage	Description of typical damage
Up to 30	Very slight or negligible	Hair cracks in plaster. Perhaps isolated slight fracture in the building, not visible on outside.
30–60	Slight	Several slight fractures showing inside the building. Doors and windows stick slightly, service to decoration probably necessary.
60–120	Appreciable	Slight fracture showing outside of building (or main fracture). Doors and windows sticking, service pipes may fracture.
120–180	Severe	Service pipes disrupted. Open fractures requiring rebonding and allowing weather into the structure. Window and door frames distorted; floors sloping noticeably; walls leaning or bulging noticeably. Some loss of bearing in beams. If compressive damage, overlapping of roof joints and lifting of brickwork with open horizontal fractures.
>180	Very severe	As above, but worse, and requiring partial or complete rebuilding. Roof and floor beams lose bearing and need shoring up. Windows broken with distortion. Severe slopes on floors. If compressive damage, severe buckling and bulging of the roof and walls.

4.5 MINE SUBSIDENCE MONITORING TECHNIQUES

Monitoring of mining induced subsidence is of vital importance to mining professionals and community for taking appropriate mitigation measures to avoid and control subsidence. Depending on the application, the monitoring techniques can be divided into two classes as direct monitoring methods and indirect monitoring methods (Figure 4.6).

4.5.1 Indirect monitoring techniques

Indirect monitoring techniques involve rock mechanics measurements, measurement of structures with tiltmeters and geophysical measurements.

Rock mechanical measurements

They mainly focus on determining stresses in the rock mass using load measurement instruments. Load cells (deformation measurement devices) are placed on the surface of the measurement hull generated in the rock mass. Load cells had first been investigated by Kelvin in 1930 and it was realized that the electrical conductivity of metallic transducers changes when they are exposed to tension. Basically, all load cells are designed to convert mechanical movement to electrical signs. The ideal load cell generates electrical signals from deformation in the load cells from external forces. Moreover, temperature, manufacture defects, amount of adhesion of the load cell to

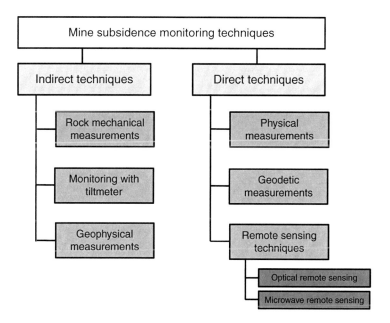

Figure 4.6 Mine subsidence monitoring techniques

the ground, and physical strength of the metal to bending impacts all change the electrical conductivity. The amount of deformation in the rock mass determined by load cells can be an indicator for subsidence monitoring.

Monitoring with tiltmeter

The tiltmeter is a sensor capable of measuring dynamic and static angular movements at a point. A portable tiltmeter is placed temporarily on a reference plane which is normally an entrance of a gallery or a tunnel or an upper rock at the surface. The speed and amount of any angular deformation at a point can be estimated by regular measurements at these reference planes. Angular deviations, resulted by subsidence, can be measured by fixed tiltmeters due to their abilities of continuous monitoring. Tiltmeters measure a slope at a point and can be mounted at the surface of structures on the ground.

Geophysical measurements

In-situ field investigations and geophysical surveys can reveal important discontinuity information which cannot be detected visually. Also, the degree of fracturing and extent of cracks in a rock mass can be represented in a relatively short time. Moreover, the rock mass quality, alteration degree, moisture content, boundaries of different formations, fault zones, fractures, and groundwater level can also be determined using geophysical measurements. Hence, geophysical measurements performed in the caving zone may give indirect information about the level of fractures to be used in the subsidence predictions.

4.5.2   Direct monitoring techniques

Direct monitoring techniques mainly consist of physical measurements, geodetic measurements, and remote sensing methods.

Physical measurements

Physical measurements are usually performed using extensometer, inclinometer, and electronic devices. An extensometer is a device that is used to measure changes in the length of an object. Its name comes from "extension-meter". The principle behind the extensometer is measuring the deviations in the conductivity of the wire, the resistance of which changes due to extension or compression of the wire. Vertical extensometers, which consist of a pipe or cable anchored at the bottom of the borehole, are used for site-specific measurements of subsidence. The principle behind using the vertical extensometers is to measure and record continuously the distance between the bottom of the borehole and the ground surface through the pipe or cable which are connected to the recorder. The pipe or cable extends from the bottom of the borehole, through the geologic strata that are susceptible to compaction, to the ground surface. An inclinometer, on the other hand, measures the deformation through the slope changes at a point. Electronic devices are also used for physical measurement purposes by measuring rock pressures, forces on rock masses, stress induced fractures, and slope alterations at the surface.

Geodetic measurements

Geodetic measurements can be classified into three groups as:

i)   Sensitive triangulation networks, sensitive polygon series and intersection points and observations based on the determination of movement at a horizontal direction.
ii) Sensitive geometric leveling measurements using automatic/digital levels and observations based on the determination of movement at a horizontal direction.
iii) Global Positioning Systems (GPS): Leveling using Global Positioning System (GPS) surveying or conventional leveling are alternatives to vertical extensometers. GPS surveying is used to monitor subsidence over greater distances or at a regional scale. Benchmarks or geodetic stations are used along a transect or network. Ground elevations at each benchmark can be obtained within plus or minus 2–3 cm of accuracy with GPS surveying. For regional scale surveys of this type, conventional leveling is less accurate.

The land surface elevations are initially surveyed and then re-surveyed every few years to track changes in elevation at the benchmarks and monitor trends over time. However, these conventional field surveying techniques have some limitations and disadvantages like (Ge *et al.*, 2007):

1   Time consuming, labor intensive and costly due to the fact that they monitor ground subsidence on a point-by-point basis.
2   Monitoring is usually constrained to much localized areas, and it is very difficult to monitor any regional deformation induced by underground mining.
3   Even in the localized areas, the monitoring points are not usually close enough to assist in understanding the mechanisms involved in ground subsidence.

In order to eliminate the deficiencies of conventional monitoring techniques, alternative remote sensing techniques are utilized in mine subsidence monitoring.

Remote sensing techniques

Subsidence measurements include surveys of surface and subsidence movements and measurements and monitoring of surface topographical changes. Dynamics of mine subsidence, as influenced by various factors, is difficult to model, hence the subsidence estimation models can highly be erroneous. Wright and Stow (1999) claimed that the reported actual to predicted subsidence results can range from 48% to 773% without adjustments. Essentially, a reliable mine subsidence monitoring technique is a key to eradicating the deficiencies of empirical subsidence estimation models. Various remote sensing methods serve for subsidence monitoring at different accuracies.

4.6   MINE SUBSIDENCE MONITORING USING REMOTE SENSING

In the last decade, the utilization of remote sensing techniques in detecting and monitoring of mine subsidence have been receiving growing attention due to the increasing power of computational methods and accessibility of the remotely sensed data (Table 4.2). Also, remote sensing technologies do not require such detailed field surveying studies as conventional techniques do. Remote sensing techniques utilized in mine monitoring can be divided into two main groups according to their types of sensors utilized as optical (passive) remote sensing and microwave (active) remote sensing. Optical remote sensing techniques use aerial photographs and satellite images acquired from optical sensors. Microwave remote sensing techniques utilize synthetic aperture radar interferometry. Both optical and microwave remote sensing techniques use the change between two digital elevation models of the same area obtained from multi-temporal remotely sensed data.

4.6.1 Optical remote sensing in mine subsidence monitoring

Mine subsidence monitoring using optical remote sensing can be achieved through aerial photographs and satellite images as input data. In the case of aerial photographs photogrammetric techniques are utilized to process the data and to generate metric and descriptive information from analog and digital images. Photogrammetric reconstruction produces surfaces that are rich in semantic information, which can clearly be recognized in the captured imagery. The inherent redundancy associated with photogrammetric restitution results in highly accurate surfaces. Nevertheless, the amount of effort and time required by the photogrammetric reconstruction procedure is a major

Table 4.2 List of remote sensing techniques used in mine subsidence monitoring

Method	References	Data Used	Case Study
SAR	Wright and Stow, 1999	ERS SAR data and SAR SLC data	Selby coal field, UK
DInSAR, GIS integration	Chang et al., 2004	ERS-1/2, JERS-1	Australia
Multi-temporal satellite images and GIS	Zhao et al., 2005	Landsat5 TM and Landsat ETM+	Underground coal mine subsidence monitoring in Tang Shan, Hebei Province, China
PSInSAR	Jung et al., 2007	25 JERS-1 SAR Interferometric pairs	Gaeun coal mining area, Korea
DInSAR	Ge et al., 2007	ERS-1/2, JERS-1, RADARSAT-1 and ENVISAT	The Tower Colliery, Southwest of Sydney, Australia
LiDAR and InSAR	Froese and Mei, 2008	LiDAR data and RADARSAT-1	Turtle Mountain, in the Crowsnest Pass, Alberta, Canada
DInSAR and GIS	Fang et al., 2008	ERS and ENVISAT images	Underground coal mine subsidence monitoring in Tang Shan, Hebei Province, China
Multi-temporal satellite images	Quanyuan et al., 2009	TM/ETM+ and SPOT5 images	Longkou city, Shandong Province, China
SBAS DInSAR	Castañeda et al., 2009	27 ERS-1/2 images	Ebro River, Spain

disadvantage (Habib *et al.*, 2007). Multi-temporal satellite images were also used to monitor the mine subsidence area dynamically through post classification comparison analysis (Zhao *et al.*, 2005). The objective of post classification comparison is to identify the topographical changes between multi-temporal image data covering the same spatial region.

4.6.2 Microwave remote sensing in mine subsidence monitoring

Interferometric Sythetic Aperture Radar (InSAR) is an embedded technique to observe the Earth's topography using different look angles to compare measurements of the same objects and it is an increasingly popular alternative to extensometers and GPS or conventional surveying methods. Interferometry refers to techniques that utilize the coherent properties of electromagnetic radiation to compare two or more waves. InSAR is a space-borne remote sensing technique that uses changes in satellite radar signals created by interferences on the Earth's surface to measure changes in land surface elevation. The InSAR approach can be single-pass and repeat-pass depending on the number of visits to the same area.

Synthetic Aperture Radar (SAR) looks at the cross-track direction (direction perpendicular to the direction of motion) or along-track direction and uses coded waveforms to obtain fine resolution in the cross-track direction while using the along-track motion to synthesize a large antenna thereby obtaining fine resolution in the along-track direction (Figure 4.7).

Accordingly, the elevation (h) can be calculated using Eq. 4.1.

$$h = H - r \times \cos\theta \tag{4.1}$$

In Eq. 4.1, altitude of the satellite (H) and distance from the satellite to the target area (r) can be determined based on the satellite orbit and attitude data. From Figure 4.7 and

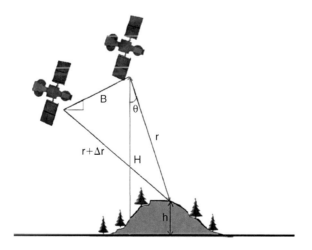

Figure 4.7 Principle of InSAR measurement

Eq. 4.1, elevation of the target (h) is calculated using θ (off-nadir) angle. Consequently, the principle of SAR interferometry relies on elevation data by calculating the off-nadir angle based on the phase difference of two SAR images. In this sense, the accuracy of elevation data depends on the accuracy of satellite altitude (H) and distance between the satellite and target of interest. Therefore, the accuracy of orbit and attitude data is crucial for the SAR interferometry processing.

The phase difference of two complex SAR images in the same area is a function of topography and ground displacement. Therefore, it can be used to measure and track subtle crustal deformations in the Earth's surface caused by earthquakes, volcanoes, and by groundwater and fossil fuel extraction and injection. Similar to GPS, InSAR enables measurement of subsidence on a regional scale and, like extensometers, the accuracy of elevation measurements with InSAR can be within a fraction of an inch. InSAR is a cost-effective means of monitoring subsidence. Another advantage of SAR interferometry is the improved subsidence model accuracy due to the vastly increased quantity of available feedback data (Wright and Stow, 1999).

InSAR has limitations of decorrelation of noise caused by random temporal variations of terrain reflectivity and atmospheric noise related to random fluctuations of atmospheric refraction (Berardino et al., 2003). It is especially difficult to apply SAR interferometry to mountainous regions with forest cover where the branches and leaves of dense vegetation cause volume scattering, resulting in severe temporal decorrelation. Hence, it is not well suited for the predominantly agricultural lands where farming affects the surface elevation of lands. For example, InSAR detects the changing surface elevation of a developing crop canopy and not the actual elevation of the land, and where land leveling is practiced. Similarly, InSAR may provide the changes in land elevations resulting from leveling and not resulting from subsidence.

Differential Interferometric Synthetic Aperture Radar (DInSAR) is a space-borne remote sensing technique that allows the detection and measurement of ground deformation and surface displacements over large areas with high spatial resolution in centimeter to millimeter accuracy (Jung et al., 2007; Castañeda et al., 2009). The technique is based on the measurement of the phase variation (interferometric phase) between successive radar acquisitions and requires the use of a digital elevation model (DEM) to remove the altitude contribution from the interferometric phase. The ground displacements are calculated on the SAR sensor's line of sight (LOS).

Differential Interferometric Synthetic Aperture Radar (DInSAR) has the ability to cover a large area of the Earth and data acquisition can be done in a relatively short time as compared to the conventional methods. SAR data has a potential to provide change information with a potential resolution of the order of one cm in the vertical direction and 20 cm in the horizontal direction (Hartl et al., 1993).

DInSAR approaches can be classified in two main categories as Persistent Scatters (PS)-like and coherence-based depending on the type of detected scatters. The PS-like approach operates on single-look interferograms, generated with respect to a reference (master) image, without any constraint on the spatial and temporal baselines of the SAR data acquisition orbits (Castañeda et al., 2009). This methodology allows the analysis of single targets that exhibit sufficiently stable radar reflectivity and are almost unaffected by temporal and spatial decorrelation. The coherence-based approaches use an appropriate combination of averaged (multi-look) differential interferograms, characterized by relatively small spatial and temporal baselines, in order to reduce the decorrelation effects and detect not only PS but also distributed scatters (Castañeda *et al.*, 2009).

4.7 CASE STUDY: MINE SUBSIDENCE MONITORING USING AERIAL PHOTOGRAPHS

Data and study area

In this section, a case study is presented to provide the reader with the methodology of utilizing remote sensing techniques in mine subsidence monitoring. This case study was conducted by Hande Yetişen as her Master of Science thesis.

The study is conducted in Cayirhan Lignite Coal Mine, in the western part of Ankara in Turkey. The mining method used in the mine is fully mechanized retreat longwall mining. The structural geology of the region includes anticlines, synclines, faults, and folds.

Various effects of ground movement (such as cracks formed parallel and perpendicular to panel advance direction) that had occurred at the surface in Çayırhan coal mine due to retreating longwall panels are presented in Figure 4.8.

In this study, the ground movements and the rate of subsidence at four different zones in the mine area are monitored using aerial photographs.

Methodology

The methodology followed in this study essentially entails five main stages: (i) Data acquisition, (ii) Scanning of aerial photos, (iii) Stereo image processing, (iv) DEM

Figure 4.8 Cracks at the surface (Yetişen, 2007) (See color plate section)

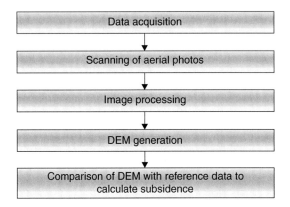

Figure 4.9 Process flowchart of the research (Yetişen, 2007)

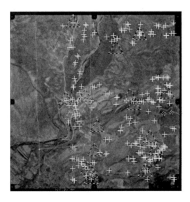

Figure 4.10 Mutual orientations of aerial photographs of the year 2001 (Yetişen, 2007)

generation, (v) Comparison of DEM with the reference data and analysis of the results. The process flow diagram is given in Figure 4.9.

Six 1/18,000 scaled stereo aerial photographs of the year 1996 and three 1/16,000 scaled stereo aerial photographs of the year 2001 were obtained from the Ministry of Environment of Turkey. The photographs of the year 1996 cover the region in two columns and the photographs of the year 2001 cover the region in one column (Figure 4.10). The aerial photographs were obtained in the form of films and were later scanned to obtain data in digital form. These photographs were scanned at 20 micron resolution using a high precision photogrammetric scanner, Vexcel Ultrascan 500 and then converted to digital form. These digital aerial photographs were then subjected to stereo image processing to obtain 3D coordinates of selected points on the images. The details of stereo image processing is explained in the case study of Chapter 5. In total 795 points were collected.

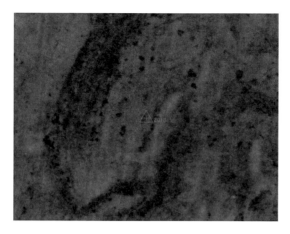

Figure 4.11 Control points measurement for balancing process (Yetişen, 2007)

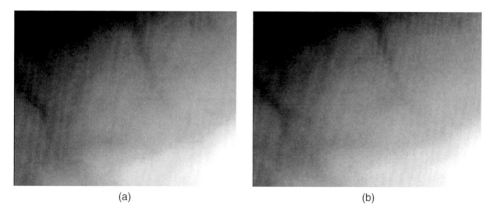

(a)   (b)

Figure 4.12 Digital elevation models of the study area for the years (a) 1996 and (b) 2001 (Yetişen, 2007)

The Ground Control Points (GCPs) used for stereo imaging are illustrated in Figure 4.11. After obtaining stereo image pairs by photogrammetric processes, the elevation values for selected points are obtained. Then using the interpolation method, DEMs of the region were obtained. The DEMs for the years 1996 and 2001 are presented in Figure 4.12. In Figure 4.12, height decreases from light color to dark color, i.e. white color shows the elevated points and black color shows the lower points in Figure 4.12.

After obtaining the DEMs of the study area for 1996 and 2001, the difference of the DEMs are taken to see the elevation difference which occurred from 1996 to 2001. Figure 4.13 and Table 4.3 represent the 3D elevation difference map and the change in the horizontal (x and y) and vertical coordinates of the selected points, respectively.

When two DEMs of the years 1996 and 2001 were overlapped, it was observed that the minimum elevation decreased from 668 m to 664 m from 1996 to 2001. This difference shows the impact of mine subsidence on the surface.

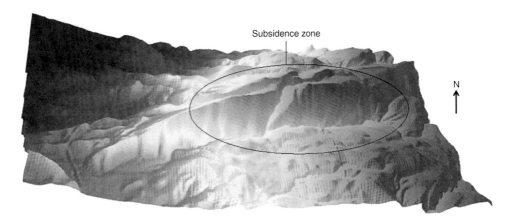

Figure 4.13 3D Difference DEM of the study area (Yetişen, 2007) (See color plate section)

Table 4.3 Subsidence monitored in the study area from 1996 to 2001 (Yetişen, 2007)

	1996			2001			1996–2001 Diff. (m)		
Point	x	y	z (m)	x	y	z (m)	dx	dy	dz
31	386132.67	4443584.44	705.25	386131.28	4443584.04	701.77	1.39	0.40	3.48
32	386604.62	4443782.40	750.03	386602.87	4443781.68	745.91	1.75	0.72	4.12
33	386649.89	4443786.83	752.29	386648.42	4443786.49	748.32	1.47	0.34	3.97
34	386640.82	4443779.52	751.67	386639.06	4443779.07	747.20	1.76	0.45	4.47
35	386992.55	4443790.77	778.91	386991.76	4443789.05	775.13	0.80	1.72	3.78
36	387204.26	4443598.73	764.42	387202.77	4443598.93	760.98	1.49	−0.20	3.45
37	387047.27	4443548.51	734.12	387046.53	4443548.24	730.21	0.74	0.26	3.91
38	387034.63	4443459.16	728.60	387033.45	4443459.39	725.01	1.19	−0.23	3.59
39	387059.73	4443440.67	732.95	387058.58	4443441.42	730.45	1.16	−0.74	2.50
40	386826.37	4443469.58	704.53	386825.25	4443469.34	700.77	1.12	0.24	3.76
41	386556.29	4443366.79	687.72	386554.84	4443367.66	684.16	1.44	−0.87	3.56
42	386039.83	4443240.23	659.54	386038.79	4443240.38	655.66	1.04	−0.15	3.88
52	386659.77	4443162.34	732.01	386659.58	4443163.33	729.77	0.19	−0.99	2.24

As is illustrated by the case study, remote sensing based on aerial photos provide an efficient tool for monitoring mine subsidence for large areas.

REFERENCES

Bell, F.G. (1975) *Site Investigations in Areas of Mining Subsidence*. London, Newnes Butterworths. 168 pages.

Berardino, F., Fornaro, G., Lanari, R. & Sansosti, E. (2002) A new algorithm for surface deformation monitoring based on small baseline differential SAR interferograms. *IEEE Transactions on Geoscience and Remote Sensing*, 40, 2375–2383.

Bruhn, R.W. & Speck, R.C. (1986) *Characteristics of subsidence over pillar extraction panels*. U.S. Bureau of Mines, Contract Report J0233920, GAI Consultants, Inc.

Castañeda, C., Gutiérrez, F., Manunta, M. & Galve, J. (2009) DInSAR measurement of ground deformation by sinkholes, mining subsidence, and landslides, Ebro River, Spain. *Earth Surface Processes and Landforms*, 34, 1562–1574.

Chang, H.C., Ge, L. & Rizos, C. (2004) Environmental impact assessment of mining subsidence by using spaceborne radar interferometry. *Third FIG Conference*, Jakarta, Indonesia, 3–7 October 2004.

Fang, M., Mingxing, Y., Xiaoying, Q., Chengming, Y., Baocun, W., Rui, L. & Jianhua, C. (2008) Application of DinSAR and GIS for underground mine subsidence monitoring, *The International Archives of the Photogrammetry, Remote Sensing and Spatial Information Sciences*, Vol. XXXVII, Part 81, Beijing.

Froese, C.R. & Mei, S. (2008) *Mapping and Monitoring Coal Mine Subsidence Using LiDAR and InSAR*, GeoEdmonton'08. Available from http://www.ags.gov.ab.ca/geohazards/pdf/tm_coal_lidar.pdf (Accessed 11th August 2010).

Ge, L., Chang, H.C. & Rizos, C. (2007) Mine subsidence monitoring using multi-source satellite SAR images. *Photogrammetric Engineering and Remote Sensing*, 73(3), 259–266.

Habib, A.F., Bang, K.I., Aldelgawy, M., Shin, W.S. & Kim, K.O. (2007) Integration of photogrammetric and lidar data in a multi primitive triangulation procedure. *ASPRS Annual Conference*, Tampa, Florida, 7–11 May 2007.

Hartl, P., Thiel, K.H., Wu, X. & Xia, Y. (1993) Practical application of SAR-interferometry; experiences made by the Institute of Navigation. *Proceedings of the 2nd ERS Symposium*, Hamburg, Germany, 11–14 October 1993 (Noordwijk: ESA Publications Division), pp. 717–722.

Ioannidis, C. & Potsiou, C. (1999) Detailed restitution and representation of the Seaward Castle of Chios. *Proceedings of XVII CIPA International Symposium*, Recife, Brazil.

Jung, H.C., Kim, S.W., Jung, H.S., Min, K.D. & Won, J.S. (2007) Satellite observation of coal mining subsidence by persistent scatterer analysis. *Engineering Geology*, 92, 1–13.

Kratzsch, H. (1983) *Mining Subsidence Engineering*. Berlin, Heidelberg, New York: Springer-Verlag.

Miao, F., Yan, M., Qi, X., Ye, C., Wang, B., Liu, R. & Chen, J. (2008) Application of DinSAR and GIS for underground mine subsidence monitoring. *The International Archives of the Photogrammetry, Remote Sensing and Spatial Information Sciences*, XXXVII (Part B1), 251–255, Beijing.

Miao, F., Yan, M., QI, X., Ye, C., Wang, B., Liu, R. & Chen, J. (2008) Application of Dinsar and GIS for underground mine subsidence monitoring. *The International Journal of Photogrammetry, Remote Sensing and Spatial Information Sciences*, 37 (Part B1), 251–255.

Quanyuan, W., Jiewu, P., Shanzhong, Q., Yiping, L., Congcong, H., Tingxiang, L. & Limei, H. (2009) Impacts of coal mining subsidence on the surface lanscape in Longkou city, Shandong province of China, *Environmental Earth Sciences*, 59, 783–791.

Shadbolt, C.H. (1977) Mining subsidence – historical review and state of art, presented to the *Conference on Large Ground Movements and Structures*, Cardiffs, Wales, July 1977.

URL 1, http://www.dep.state.pa.us/dep/deputate/minres/Districts/homepage/California/Underground/Mine%20Subsidence/mine_subsidence.htm (Accessed 9 February 2010).

Wright, P.L. & Hoffman, G.L. (1995) *Measurement and Control of Mining Subsidence: A Handbook for Western Canada*. Devon, Alberta, Canada, The Coal Mining Research Company. 223 pages.

Wright, P. & Stow, R. (1999) Detecting mining subsidence from space. *International Journal of Remote Sensing*, 20(6), 1183–1188.

Yetişen, H. 2007. M.Sc. Thesis, Identification and Assessment of Subsidence Using Aerial Photogrammetry, Hacettepe University, Ankara, Turkey.

Zhao, H., Guo, Q. & Wang, S. (2005) The application of remote sensing and GIS on coal mining subsidence area. *IEEE*, 1550–1553.

Chapter 5

Remote sensing in slope stability monitoring

This chapter briefly covers the use of remote sensing in slope monitoring. It discusses the causes and consequences of slope failures in surface mines and it emphasizes the significance of slope monitoring for effective mining. The typical use of remote sensing is presented by a case study performed for Ovacık abandoned coal mine in Turkey.

5.1 AN OVERVIEW

Slope stability in open pit mines is of paramount concern in mining engineering. The consequences of slope failure (Figure 5.1) include deferred production, machine and equipment loss, and even loss of human life. Therefore, sustaining safe pit slope angles and monitoring the stability of the pit slopes through time are vital to technically and economically effective mining. The conventional slope stability monitoring techniques like surveying with topographic measurements may not be cost and time effective. In this regard, remote sensing techniques can be considered as an alternative slope monitoring technique. High resolution satellite data integrated with GIS is a valuable

Figure 5.1 Slope failure views from an open pit copper mine (See color plate section)

tool in determining in advance any displacement on the slope due to failure. This chapter presents the use of remotely sensed data and remote sensing in slope stability monitoring.

Although there are various failure mechanisms, such as circular failure, toppling, planar failure, etc., pit slope failures usually occur in the form of sliding along single or multiple discontinuities, where the geometry of the rock mass behind the slope face plays a critical role in the stability of slopes (Figure 5.1). The geometrical relationship between the discontinuities in the rock mass and the designed slope determines whether parts of the rock mass are free to slide or fall. In this respect, there are various factors playing an important role in the slope failure mechanism: the shear strength of the potential failure surface, discontinuities of the rock mass, pit geometry including overall pit slope angle and pit depth, water presence, and dynamic forces.

In addition to open pit slopes, slope instability problems may occur in the dump sites, stock piles of the ore and in tailing dams, where the slope material has mainly fragmented and/or is soil-like in nature. In this case, the frictional properties of the material, as well as the geometry of the piles or dam and existence of water plays a critical role in slope stability.

5.2   PROBLEMS WITH SLOPE INSTABILITY

Slope failures in surface mining may result in various undesired outcomes. In large-scale mines with high benches, the failure of a steep slope causes sliding of a substantial amount of material into the operating faces and haulage roads, and hence blocking the operations. As a result, it causes deferred production and significant losses associated with it. In order to reactivate mining operations and material haulage, this mass should be removed, which may not be possible in some cases. Mass removal due to slope failure requires allocation of machinery and equipment, which are originally used for production and material haulage, for a long period of time. Spending scheduled operating hours for mass removal increases operating and maintenance costs. The undesired impacts of slope failure can even make the mining project infeasible. In some extreme cases, machine and equipment loss, injuries and loss of human life could also happen. Figure 5.2 illustrates a rural settlement endangered by the slope instability of a pit which is abandoned without proper closure measures.

Design of open pit slope angles is becoming more and more important as the depths of open pits continuously increase. Small changes in the overall pit slope angle have large consequences on the overall economy of the mining operation. A case in particular is the Aitik open pit mine in the northern Sweden, which currently faces the design of the overall slope angles for continued mining toward a depth of around 500 meters. From the collection of a number of case studies from North and South America, Africa, Asia, and Europe, several examples of large-scale failures were found, especially in weak rocks. There are much fewer examples of slope failures in hard and brittle rocks where slope instability is governed by the discontinuity sets. The few cases found indicate that a failure in this type of rock is more uncontrollable.

In order to emphasize the significance of slope failure in terms of mine economics an example from Zambia Copper Mine can be considered. A major slope failure occurred

Figure 5.2 Slope failure of an abandoned pit close by residential houses

in the south face of the Nchanga Open Pit on Sunday, April 8, 2001. Over 5 million tones of failed material collapsed into the pit. Ten employees, who were working in the bottom of the pit at the time of failure, lost their lives in this tragic accident. The operations were severely affected. Several pieces of major equipment were destroyed in the accident and had to be replaced. A full-scale rescue and recovery operation was launched as soon as conditions were established. Consequently, the mine production was decreased by 21%.

Similar to pit instability, the slope failures in dump sites, stock piles and tailing dams can have detrimental consequences in various ways. Dump site and stock pile failures can result in temporary/permanent blockage of the site during mining operations. As in the case of mine closure, the stability of pit slope as well as slopes of the dump site should be provided for safe relinquishment. Slope failures occurring in the tailing dams results in flooding of the downstream environment as well as polluting it. As a result, any slope failure in surface mines should be avoided for safe and economic mine operation and closure.

There are several ways to reduce the risks of slope failures and to enhance the overall rock mass strength, such as safe geotechnical designs, supports, rock fall catchment systems, monitoring devices for advance warning of impending failures, and efficient draining of slopes which should be implemented. Diligent monitoring and examination of slopes for failure is one of the most important means of protecting the mine environment. Even the most carefully designed slopes may experience failure from unknown geologic structures, unexpected weather patterns or seismic shocks.

5.3 MEASUREMENT, INSTRUMENTATION, AND MONITORING SYSTEMS

In order to control and mitigate the risk of slope failure, continuous slope monitoring is essential. Conventional field measurement techniques and instrumentation (explained in Chapter 4 for mine subsidence) may not practically be used for large-scale mines. These costly and labor intensive techniques could also be error prone due to the limited number of data collected for a particular region. In these cases, monitoring the slopes of a surface mine through remote sensing could be a solution. Slope stability monitoring with the help of remote sensing mainly relies on extracting the Digital Elevation Model (DEM) of slopes for several temporal intervals. The displacements due to slope insta-bility are assessed by finding the difference between DEM in different time periods. As explained in Chapter 3, DEM extraction for slope stability monitoring in surface mines can be performed using optical, radar, and LIDAR remote sensing technologies. The choice of remote sensing system depends on the amount of displacement, degree of desired accuracy, required frequency of time periods, availability of remote sensing data, and costs. If the DEM extracted from the remote sensing data does not have accuracy compatible with the observed displacements in the mine, it cannot be used in slope stability monitoring. Due to the temporal resolution of the satellites, the desired temporal frequencies may not be met for slope monitoring. Therefore, selection of an appropriate remote sensing system for slope stability involves careful investigations of the nature of the slope instability.

5.4 A CASE STUDY: ASSESSING AND MONITORING UNSTABLE SLOPES USING REMOTE SENSING

5.4.1 Study area

This case study is presented with the courtesy of Emil (2010). The study area is an aban-doned surface coal mine site in Ovacık, Yapraklı, Çankırı which is located in Turkey about 140 km northeast of Ankara. The location of the mine site is given in Figure 5.3. Rough terrain is observed in topography with an average elevation of 1300 m above sea level. The climate of the area is terrestrial with a mean annual precipitation of 394 kg/m². The main land use in the area is forest with pine and oak trees. There are also small agricultural areas around the villages. Bare lands also exist in the area. The mining induced land use involves open pit excavation, dump sites (both afforested and not afforested) and remnants of coal stock sites, as well as demolished and abandoned administrative buildings. The current abandoned mine layout can be seen in Figure 5.4.

The coal seam operated in the Ovacık coal mine has 0.3–0.8 m thickness with a 75–80 degree inclination. The calorific value of the coal has been determined as 4500–5000 kcal/kg. The open pit void was left as it is, after the operation stopped without any stability or safety measures. There are high slopes around the pit. The pit is filled with water due to surface and underground water discharges (Figure 5.5).

5.4.2 Data acquisition

For mapping the slope instability problems in the abandoned mine site, the original topography of the site before mining and current topography of the site was obtained.

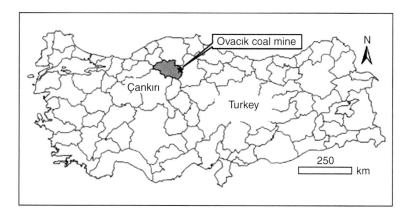

Figure 5.3 Location of study area (Emil, 2010)

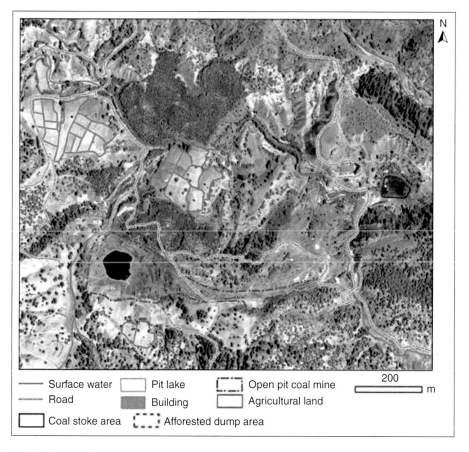

Figure 5.4 Current Ovacık coal mine layout overlaid with Worldview-1 satellite imagery (Emil, 2010)
(See color plate section)

Figure 5.5 A scene from abandoned open pit with surrounding steep slopes (Emil, 2010) (See color plate section)

DEM for pre-mining was obtained from stereo aerial photos, while the DEM for the current site was acquired from 3D Terrestrial Laser Scanner (TLS) data. Then the natural slope angles that are stable from the pre-mining DEM were determined. Finally, the unstable slopes, which are greater than the natural stable slopes, were indentified when the DEM for current site were mapped. The data set used in the case study includes aerial photographs, satellite imagery of Worldview-1, and Ground Control Points (GCP) obtained by GPS survey.

3D Terrestrial laser scan survey

A three dimensional (3D) Terrestrial laser scan survey was conducted using the instrument Optech ILRIS-3D laser scanner on the site shown in Figure 5.6. The instrument consists of a scanner, a tilt unit, a tripod, a control unit (pocket pc), and a power supply. Table 5.1 shows the specifications of the Optech ILRIS 3D Terrestrial laser scanner. The parts of instrument are shown from top to bottom as the scanner, the tilt unit, the tripod, the control unit, and the power supply in Figure 5.6. The result of scan are millions of points known as "point clouds", which are 3D (X, Y, Z) representation of the scanned terrain with respect to the scanner position.

The scanning of the site was completed in five days of field work with 12 scan schedules. The point clouds obtained by 12 scan locations were merged and geo-referenced. An oblique display of point clouds is given in Figure 5.7 showing the open pit from west and surface features such as trees. The processing of this point cloud data was computationally intensive. Data were cleaned of noises. The noisy data mainly consists of erroneous points above the terrain and the points recorded below water in

Figure 5.6 Optech ILRIS-3D laser scanner during scanning near dump site (Emil, 2010)

Table 5.1 Specification of the Optech ILRIS 3D terrestrial laser sanner (Emil, 2010)

Parameters	Unit	Optech ILRIS 3D
Scanning range (80% target reflectivity)	m	1,500
Data sampling rate	point/second	2,500
Beam divergence	degree	0.00974
Minimum spot step (X and Y axis)	degree	0.00115
Raw range accuracy at 100 m	mm	7
Raw positional accuracy at 100 m	mm	8
Laser wavelength	nm	1,500
Laser class		Class 1 (eye safe)
Scanner field of view	degree	40 × 40

the pit lake. The point cloud is left with 157 million points after noise reduction with average point spacing of 14 cm.

The terrestrial laser scan survey was conducted to gain current topography of the site in detail. Point cloud recorded in scan survey was used to produce current DEM to detect slope instability comparing with pre-mining slopes.

Global Positioning System (GPS) survey

Global Positioning System (GPS) receivers were utilized depending on device availability and accuracy level needed for the ongoing purpose of data collection. Table 5.2

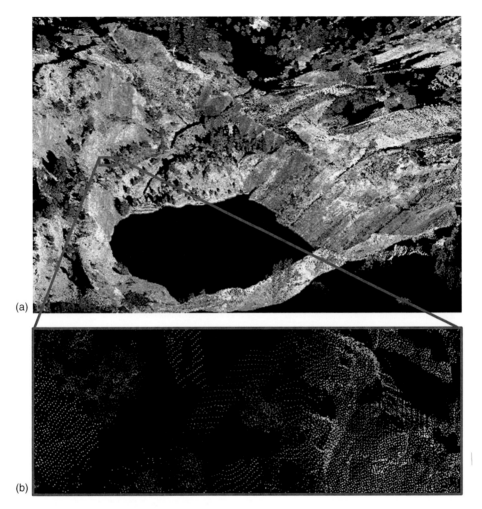

Figure 5.7 An oblique view of point cloud showing open pit looking from west (a) and close view of point cloud (b) (Emil, 2010)

shows GPS instruments used with basic specifications. In this study, a Topcon GR3 receiver was used in real time kinematic mode which gives the highest accuracy (centimeter level) by employing both the reference station correction and carrier phase signals from GPS satellites (Blanco *et al.*, 2009). On the other hand, Topcon GRS-1 and Septentrio AsteRx1 GPS receiver were used in DGPS mode which have relatively lower accuracy (tens of centimeters) by employing differential correction from a static reference station with known fixed position (Blanco *et al.*, 2009).

GPS survey was conducted in the field to collect Ground Control Points (GCPs) shown in Figure 5.8 to be used for four purposes. The first was to georeference point cloud data acquired by Terrestrial Laser Scanner (TLS). The second was to use in orthorectification of aerial photos and Worldview-1 satellite imagery. The third was

Table 5.2 Specifications of GPS receivers utilized in this study (Emil, 2010)

Type GPS Receivers	Signal Tracking	Indicated Accuracies		
		Static	Real Time Kinematic	DGPS
TOPCON GR3	GPS, GLONASS, Galileo (72 channel)	Horizontal: 3 mm + 0.5 ppm Vertical: 5 mm + 0.5 ppm	Horizontal: 10 mm + 1 ppm Vertical: 15 mm + 1 ppm	<50 cm
TOPCON GRS-1	GPS, GLONASS (72 channel)	Horizontal: 3 mm + 0.8 ppm Vertical: 4 mm + 1.0 ppm	Horizontal: 10 mm + 1 ppm Vertical: 15 mm + 1 ppm	50 cm
Septentrio AsteRx1 PRO	GPS, Galileo (24 channel)		Horizontal: 20 cm + 1 ppm Vertical: 35 cm + 1 ppm	Horizontal: 50 cm Vertical: 90 cm

Figure 5.8 The distribution of GCPs recorded by GPS receivers (Emil, 2010)

to assess the accuracy of DEMs created by aerial photos and TLS data and the last was to use for mapping purposes. The number and usage of GCPs according to GPS receiver and date of the field work are given in Table 5.3.

Aerial photographs

Historical aerial photo sets were obtained in digital format after being scanned from the General Command of Mapping to obtain topography of mine site before mining operations. The flights took place in 1951, 1971, and 1990, capturing black and white air photos with scales of 1:35,000, 1:20,000, and 1:40,000, respectively. Also, the coverage of air photos with air photo centers are displayed in Figure 5.9. Photo boundaries given in Figure 5.9 were drawn using photo corner coordinates obtained

Table 5.3 Number of GCPs collected in the field works (Emil, 2010)

| | Date of GPS Survey and GPS Receivers Used | | | |
| | 04.07.2009 | 11.10.2009 | 15.04.2010 | |
Usage of GCPs	Septentrio AsteRx I	Topcon GRS-1	Topcon GR3	Total
Georeferencing of TLS Data	–	47	84	131
Orthorectification of Aerial Photos	–	–	6	6
Orthorectification of Worldview-1	–	–	6	6
Accuracy Assessment of Current DEM	–	–	70	70
Accuracy Assessment of Past DEM	–	82	41	123
Mapping	51	–	–	51

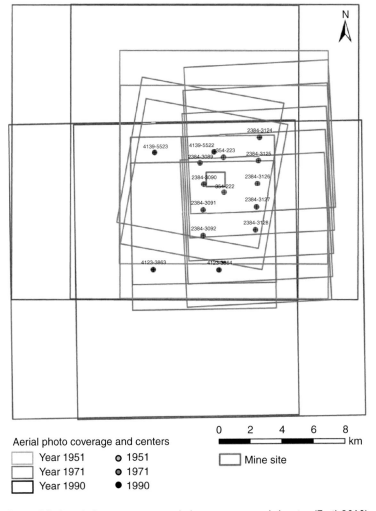

Aerial photo coverage and centers

Year 1951	○ 1951
Year 1971	○ 1971
Year 1990	● 1990

0 2 4 6 8 km

☐ Mine site

Figure 5.9 Aerial photo coverage and photo centers and the site (Emil, 2010)

from the General Command of Mapping. As seen in Figure 5.9, between photo pairs in the same flight line, there are 60% overlaps allowing to create stereo models. The stereo aerial photo pair captured in 1951 was selected for the extraction of pre-mining topography.

5.4.3   Data processing

Data processing steps refers to all pre-processing steps to generate the dataset to be used in slope instability analysis. Main data processes were the generation of Digital Elevation Models (DEM) for pre- and post-mining stages of the area and orthorectification of historical aerial photos and Worldview-1 satellite imagery.

Digital elevation model (DEM) generation for pre-mining
terrain from stereo pair of aerial photos

A DEM is a raster image storing elevation values of a surface terrain in each pixel. A stereo pair of aerial photos containing rational polynomial coefficients (RPCs) were used to generate DEM for the year 1951 in which there was no mining activity in the study area. RPCs were required to generate and orthorectify DEM from scanned stereo aerial photos taken by a frame camera. RPC's are calculated representing sensor geometry in which object point, perspective center, and image point are all on the same space using the collinearity equation. The technique consists of transformations involving pixel, camera, image space, and coordinate systems. The ENVI 4.7 *Build RPCs* tool was used to compute and include RPCs to the header files of stereo aerial photo pairs taken in 1951. In order to compute RPCs, interior and exterior orientation parameters should be calculated. Interior orientation (sensor geometry) parameters transform the pixel coordinate system to the camera coordinate system. On the other hand, exterior orientation parameters determine the position and angular orientation associated with the image. Principal point offsets, focal length and positions of the fiducial marks were required to compute interior orientation parameters. The exterior orientation parameters were obtained from GCPs which have six transformation parameters such as projection center coordinates (X, Y, and Z) and three rotation angles (omega, phi, and kappa) shown in Figure 5.10.

The camera used for the first flight was Carl Zeiss RMK 10/18. The associated calibration file was also obtained from the General Command of Mapping. The file

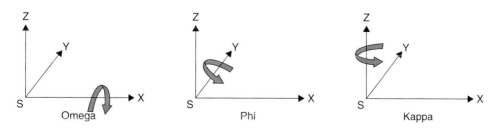

Figure 5.10 Graphical representation of rotation angles (Emil, 2010)

Table 5.4 Camera calibration parameters used in calculation of RPCs for air photo taken in 1951 (Emil, 2010)

Calibration Parameters		Value (mm)	
		99.66	
Focal Length	#	X	Y*
Principal Point Coordinates		0.015	0.005
Fiducial Point Coordinates	1	0	87.985
	2	87.995	0
	3	0	−87.985
	4	−87.995	0

*Y is in the flight direction

Table 5.5 Error calculation in fiducial point selection of the aerial photo 354-222 (Emil, 2010)

	Fiducial Points		Coordinates in pixels						
	Fiducial (mm)		Image		Predicted		Error		
#	x	y	x	y	x	y	x	y	RMS Error
1	0.00	87.98	8475.03	4396.00	8474.51	4396.51	−0.52	0.51	0.72
2	88.00	0.00	4395.97	8471.00	4396.48	8470.49	0.51	−0.51	0.72
3	0.00	−87.98	320.00	4397.96	319.48	4398.47	−0.52	0.51	0.72
4	−88.00	0.00	4397.00	325.00	4397.52	324.49	0.52	−0.51	0.72
							Total RMS Error		0.72

included data about any potential errors for the camera and the parameters displayed in Table 5.4 which are input into the photogrammetry module of ENVI 4.7.

Fiducial marks were selected from the images and corresponding fiducial coordinates given in Table 5.4 were introduced into the software to compute the interior orientation of the aerial photos. Associated image coordinates and error values of fiducial points are given in Table 5.5 and Table 5.6 for the air photos 354-222 and 354-223, respectively. As seen in Table 5.5 and Table 5.6, RMS errors of fiducial points were found acceptable (below 1 pixel) as 0.72 and 0.52 pixels for the air photos 354-222 and 354-223, respectively.

During computation of exterior orientation parameters among discernable GCPs from the aerial photos, 6 GCPs recorded by Topcon GR3 GPS receiver, which were well spread across the images, were selected (Figure 5.11). The same GCPs were used for both images. Image coordinates and error values of these GCPs are given in Table 5.8 and 5.9, for the images 354-222 and 354-223, respectively. Final RMS errors were found to be 18.48 and 12.90 pixels for the images 354-222 and 354-223, respectively.

Calculated exterior orientation parameters are given in Table 5.9 and they were used to calculate RPCs. RPCs were calculated using interior and exterior orientation

Table 5.6 Error calculation in fiducial point selection of the aerial photo 354-223 (Emil, 2010)

#	Fiducial Points Fiducial (mm) x	y	Coordinates in pixels Image x	y	Predicted x	y	Error x	y	RMS Error
1	0.00	87.98	8529.00	4525.00	8528.77	4525.50	−0.23	0.50	0.55
2	88.00	0.00	4397.00	8549.00	4397.23	8548.50	0.23	−0.50	0.55
3	0.00	−87.98	370.97	4421.97	370.74	4422.47	−0.23	0.50	0.55
4	−88.00	0.00	4502.04	399.96	4502.27	399.46	0.23	−0.50	0.55
							Total RMS Error		0.55

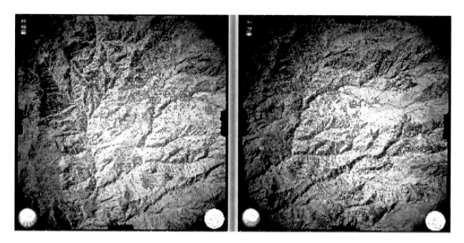

Figure 5.11 Location of GCPs on stereo aerial photos 354-222 (left) and 354-223 (right) (Emil, 2010)

parameters for each image and added to corresponding header files for the image 354-222. After computation of RPCs, it was possible to determine ground resolution of scanned aerial photos. Therefore, pixel sizes were found as 0.73×0.73 m.

Digital elevation model (DEM) generation from aerial photos

A digital elevation model (DEM) was generated using a stereo pair of aerial photos having RPCs for the year 1951 in which there was no mining activity in the study area. In the DEM extraction process from aerial photography, the *DEM Extraction Module* of ENVI 4.7 was used. There are three main steps of the DEM extraction procedure: epipolar image generation, image matching, and DEM geocoding. The implemented DEM extraction tool has the processing steps given in Figure 5.12.

Epipolar image generation needs tie points to define the relationship between the pixels in the stereo pair. An epipolar image is a stereo pair in which the left and right images are oriented in such a way that ground feature points have the same

Table 5.7 Error calculation in GCP selection of the aerial photo 354-222 (Emil, 2010)

| | Map (m) | | | Coordinates in pixels | | | | | | RMS Error |
| | | | | Image | | Predicted | | Error | | |
#	x	y	z	x	y	x	y	x	y	
1	590669.62	4514250.11	1254.70	6997.29	3341.86	7025.16	3350.61	27.87	8.75	29.21
2	590939.44	4513258.09	1233.38	5681.57	3458.43	5658.41	3447.44	−23.16	−10.99	25.63
3	590707.94	4513630.15	1262.10	6222.14	3244.57	6211.50	3245.24	−10.64	0.67	10.66
4	590959.54	4514999.29	1331.70	8000.50	3881.50	7987.51	3876.52	−12.99	−4.98	13.91
5	593650.89	4513568.91	1241.20	5396.50	6909.50	5402.21	6911.70	5.71	2.20	6.12
6	590170.30	4513042.96	1293.65	5608.50	2409.50	2413.85	2413.85	13.21	4.35	13.91

Total RMS Error 18.48

Table 5.8 Error calculation in GCP selection of the aerial photo 354-223 (Emil, 2010)

| | Map (m) | | | Coordinates in pixels | | | | | | RMS Error |
| | | | | Image | | Predicted | | Error | | |
#	x	y	z	x	y	x	y	x	y	
1	590669.62	4514250.11	1254.70	3853.50	3085.75	3873.88	3090.43	20.38	4.68	20.91
2	590939.44	4513258.09	1233.38	2564.00	3167.00	2549.02	3160.98	−14.98	−6.02	16.15
3	590707.94	4513630.15	1262.10	3089.75	2968.00	3079.86	2968.51	−9.89	0.51	9.90
4	590959.54	4514999.29	1331.70	4753.50	3627.75	4744.47	3625.05	−9.03	−2.70	9.42
5	593650.89	4513568.91	1241.20	2228.75	6609.25	2232.71	6610.44	3.96	1.19	4.14
6	590170.30	4513042.96	1293.65	2467.75	2118.25	2477.31	2120.59	9.56	2.34	9.84

Total RMS Error 12.90

Table 5.9 Exterior orientation parameters for the stereo aerial photo pair taken in 1951 (Emil, 2010)

| Exterior orientation parameters | Aerial Photos | |
	354-222	354-223
x (m)	591849.95	591579.24
y (m)	4512552.70	4515026.77
z (m)	4905.38	4818.47
Omega (degree)	−2.146	−1.740
Phi (degree)	−0.187	−0.734
Kappa (degree)	0.425	10.840

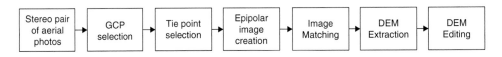

Figure 5.12 Processing steps of DEM generation from stereo aerial photos (Emil, 2010)

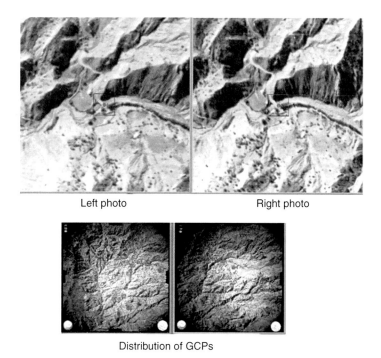

Left photo   Right photo

Distribution of GCPs

Figure 5.13 Tie point selection (Emil, 2010)

y-coordinates on both images. Epipolar images are also used in 3D visualization using anaglyph glasses.

In order to obtain absolute DEM, which is comparable with the DEM created from the point cloud measured by the Terrestrial Laser Scanner (TLS), it was necessary to use Ground Control Points (GCPs) collected in the field. GCPs tie horizontal and vertical reference systems to geodetic coordinates and then the resultant DEM is called absolute DEM. Six GCPs previously used in calculation of RPCs were also used in DEM extraction procedure. As seen in Figure 5.13, GCPs were marked in both left (354-222) and right (354-223) images and corresponding ground coordinates were introduced to match with associated image coordinates of GCPs.

Tie point selection was performed iteratively considering two main objectives: spreading tie points well across the images and reducing maximum Y parallax below 10 pixels which was the threshold value for the software. Y parallax is defined as the difference in perpendicular distances between two images of a point from the vertical plane (i.e. the air base). Due to the fact that increasing the number of tie points also increases the maximum Y parallax and makes it impossible to go further in the DEM extraction procedure, the number of tie points were kept at nine which was also the minimum number of tie points needed by the software. After reaching nine tie points supplying a maximum Y parallax of 0.3747 and spreading well across the stereo pair, tie point selections were finalized. Locations of these tie points are displayed in Figure 5.13. Image coordinates of these stereo tie points are given in Table 5.10.

Table 5.10 Stereo tie points image coordinates for image pair taken in 1951 (Emil, 2010)

| | Image Coordinates of tie points | | | |
| | Left Image (354-222) | | Right Image (354-223) | |
#	x	y	x	y
1	5026.2	2813.0	1852.7	2500.8
2	5026.8	6920.0	1899.0	6625.8
3	6946.7	7092.5	3787.2	6786.5
4	8041.0	4574.2	4779.8	4304.0
5	5025.5	4585.5	1907.0	4285.5
6	7465.5	3070.3	4271.0	2822.7
7	4290.5	3291.8	1066.5	2956.8
8	4727.2	6001.5	1606.7	5705.5
9	6840.7	4904.5	3645.8	4618.2

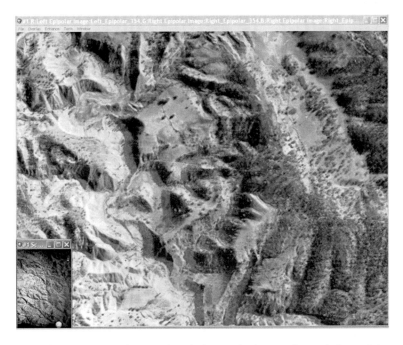

Figure 5.14 Epipolar stereo pair of images (anaglyph image) of research area before mining operation has been started (should be seen by anaglyph glasses for 3D viewing) (Emil, 2010)

Using tie points the epipolar images were produced. After loading the left epipolar image to the Green and Blue channel and the right epipolar image to the Red channel an RGB display is obtained, as seen in the Figure 5.14. The terrain in Figure 5.14 can be visualized in 3D with anaglyph glasses and it also serves for making 3D measurements of selected points in the site.

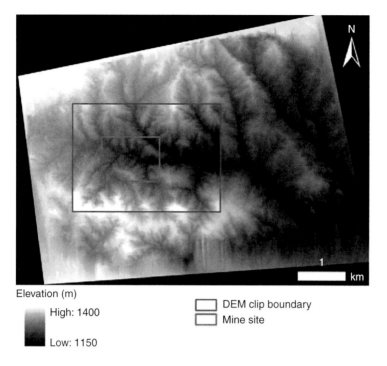

Elevation (m)

High: 1400

Low: 1150

☐ DEM clip boundary
☐ Mine site

Figure 5.15 Full scene DEM showing artifacts near the edges of the DEM and showing selected DEM boundary (Emil, 2010)

Image matching is performed prior to the DEM generation from stereo pair of images. The image matching algorithm finds the conjugate points on stereo images which belong to the same ground feature and produce a parallax image in which x-coordinate differences are used to calculate elevation values.

DEM was created with pixel sizes of 0.5×0.5 m using epipolar images obtained from stereo aerial photo pair taken in 1951. As seen in Figure 5.15, artifacts were formed in the DEM near the edges due to damages in the film of the scanned photo itself. Hence, an effective area of DEM, displayed in Figure 5.15, was clipped considering both these artifacts and the coverage of the mine site. To remove peaks in DEM caused by trees and manmade structures, median and low-pass filters of DEM were investigated with kernel sizes of 5×5, 11×11, and 21×21. After visual interpretation of hill shade views of these filtering results of DEM shown in Figure 5.16, it was decided to use the DEM edited by a low-pass filter with kernel size of 21×21 in the remaining part of this study.

Accuracy assessment of pre-mining Digital Elevation Model (DEM)

Accuracy assessment of DEM dated 1951 created by stereo pair of aerial photos representing pre-mining terrain were conducted using 123 Ground Control Points

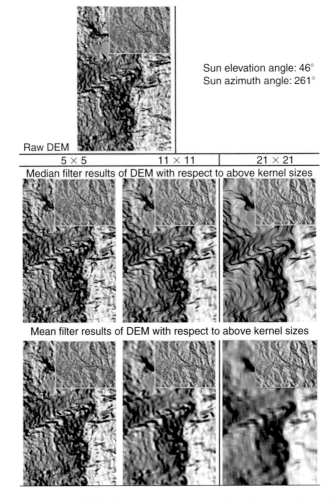

Sun elevation angle: 46°
Sun azimuth angle: 261°

Raw DEM

| 5 × 5 | 11 × 11 | 21 × 21 |

Median filter results of DEM with respect to above kernel sizes

Mean filter results of DEM with respect to above kernel sizes

Figure 5.16 Hill shade views of DEMs in the area after smoothing filters applied with kernel sizes of 5 × 5, 11 × 11, and 21 × 21 pixels (Emil, 2010)

(GCPs) shown in Figure 5.17. Because there were 58 years between data acquisition date of DEM and GPS survey, GCPs were carefully chosen from locations where no apparent topographical changes exist. Thus, only 41 of GCPs among the points recorded by Topcon GR3 could be used and others were eliminated from data in the accuracy assessment of past DEM. Instead of these eliminated points, 82 points recorded by Topcon GRS-1 were selected where there is no topographical change observed and added to the accuracy assessment. These test data were superimposed on the DEM and based on differences between Z-coordinates of these GCPs and DEM, Root Mean Square (RMS) errors were calculated. The RMS errors calculated using only Topcon GR3 and Topcon GRS-1 datasets were 4.01 m and 2.24 m, respectively. Overall RMS error was 2.95 m.

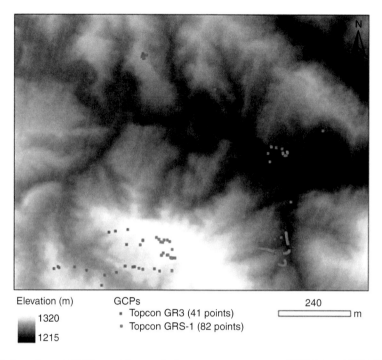

Elevation (m)   GCPs
                     ▪ Topcon GR3 (41 points)
1320   ▪ Topcon GRS-1 (82 points)
1215

240
⊏━━━━━━━━⊐ m

Figure 5.17 Locations of GCPs used for vertical accuracy assessment of DEM dated 1951 (Emil, 2010)

Orthorectification of aerial photos

Orthorectification is a process of using a DEM to correct relief displacements caused by camera tilt or terrain relief during the acquisition of aerial photos or satellite images. The orthorectification result of an aerial photo is called an orthophoto from which maps can be drawn and distances can be measured.

Orthorectification tool of ENVI 4.7 software was used in the mode of *RPC Orthorectification* for the aerial photos. DEM created by stereo pairs of images and GCPs were also used to enhance the accuracy of the orthorectification. During orthorectification nearest neighbor, bilinear, and cubic convolution interpolation methods were experimented with. As seen in Figure 5.18, nearest neighbor method gave a disjointed appearance. Also, bilinear interpolation resulted in blurring. The cubic convolution interpolation result was found to be optimal due to being sharper than bilinear and less disjointed than nearest neighbor methods. During the procedure pixel sizes were resampled to 0.5 × 0.5 m for having the same spatial resolution with Worldview-1 imagery. The resultant orthophoto created from aerial photo 354-222 is displayed in Figure 5.19 and 5.20.

Digital Elevation Model (DEM) generation for post-mining terrain from Terrestrial Laser Scanner (TLS) point cloud

Once Terrestrial Laser Scanner (TLS) surveys were merged and georeferenced, noise and points representing trees and manmade structures like buildings and electricity

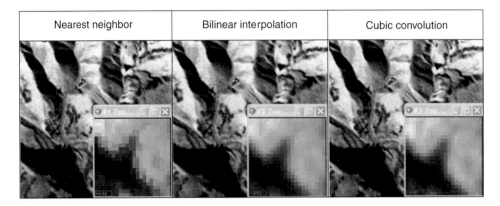

Figure 5.18 Results of resampling methods used during orthorectification of aerial photos displayed in an area of 200 × 200 m with a 10 × 10 m zoom screen (Emil, 2010)

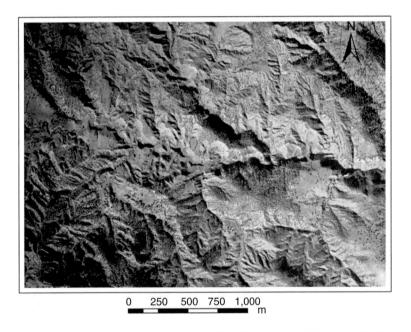

Figure 5.19 Orthorectified aerial photo (354-222) taken in 1951 (Emil, 2010)

transmission lines were eliminated. To obtain a simplified 3D model, which could easily be processed during triangulation, a data reduction was applied. After the data reduction, the number of points reduced from approximately 150 million to 5 million and average point spacing increased from 14 cm to 74 cm, respectively. The point cloud density was also calculated in units of number of points per square meter (pts/m^2). As seen from the point cloud density map given in Figure 5.21, point cloud density was in the range of 10–20 pts/m^2 which provides enough topographical detail.

Figure 5.20 Perspective view of orthophoto dated 1951 draped on DEM highlighting the mine site (Emil, 2010)

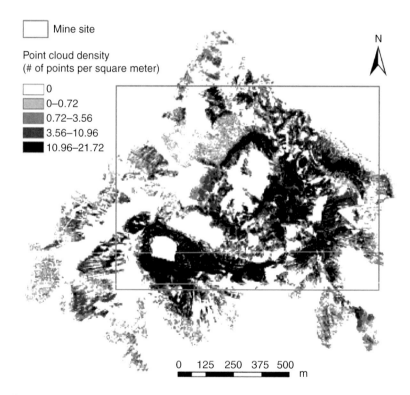

Figure 5.21 TLS point cloud density map classified to 5 classes with 5 × 5 m cell size (Emil, 2010)

The resultant point cloud was processed in *ArcGIS* 9.3 software (GIS software) to obtain DEM for the post-mining landscape (Figures 5.22 and 5.23). DEM was generated with 0.5 × 0.5 m pixels, the same size as the DEM created from stereo aerial photos.

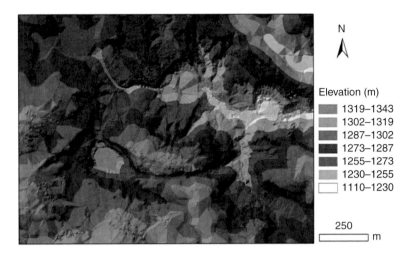

Figure 5.22 Terrain data set showing elevation (Emil, 2010) (See color plate section)

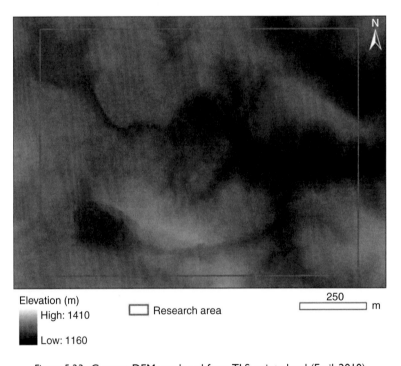

Figure 5.23 Current DEM produced from TLS point cloud (Emil, 2010)

Accuracy assessment of post-mining Digital Elevation Model (DEM)

Accuracy assessment of DEM representing current terrain is performed using 70 GCPs shown in Figure 5.24 recorded by the Topcon GR3 GPS device. 16 points were

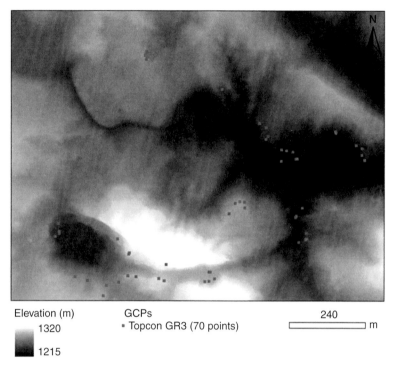

Elevation (m)   GCPs   240
    1320   • Topcon GR3 (70 points)   ⊏━━━━━━━⊐ m
    1215

Figure 5.24 Locations of GCPs used for vertical accuracy assessment of current DEM of the mine site (Emil, 2010)

eliminated from data where the point cloud density was quite low. After calculating differences between the Z-coordinates of these check points and the DEM, RMS errors were calculated. The overall RMS error was found to be 0.98 m using 70 GCPs recorded by the Topcon GR3 GPS device.

Orthorectification of Worldview-1 satellite imagery

ENVI 4.7 *rigorous orthorectification module* was used in the orthorectification process for Worldview-1 imagery which was delivered with RPCs file. GCPs, which are used for processing of aerial photos, and DEM for the current terrain were also used to enhance the accuracy of the orthorectification.

The extent of the DEM produced from the TLS was not enough to orthorectify the full scene Worldview-1 imagery of 5 × 5 km extent. Thus, it was necessary to expand the coverage of DEM by adding points shown in Figure 5.25 to the terrain dataset derived from the topographical map obtained from the General Command of Mapping to orthorectify the full scene Worldview-1 satellite imagery. During orthorectification cubic convolution resampling method was applied and pixel size remained constant at 0.5 × 0.5 m. Figure 5.26 shows the resultant orthorectified image rendered on the DEM.

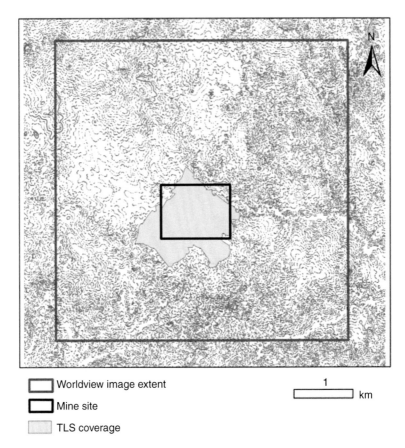

☐ Worldview image extent

☐ Mine site

▨ TLS coverage

⊢————⊣ 1 km

Figure 5.25 Elevation points derived from topographical map to expand DEM to be used in orthorectification of Worldview-1 satellite imagery (Emil, 2010)

5.4.4 Detection of unstable slopes

The slope maps were extracted using DEMs by the Spatial Analyst tool in ArcGIS 9.3 (GIS software) for the before- and after-mining stages of the mine site. As a result of mining activity in the mine site, slopes were greatly changed. The maximum angle of slopes were 67 and 84 degrees for the before- and after-mining stages of the region, respectively. Slope maps were reclassified into five classes and associated land areas were calculated for before- and after-mining terrain for each slope class. From the years 1951 to 2009, the area of each slope class larger than 30 degrees increased as a result of mining activity (Table 5.11). The steepest slope regions depicted as very steep slope (greater than 60 degrees) had been increased from 265 m^2 to 3,192 m^2.

As seen in Figure 5.27, in the classified slope map for the year 1951, slopes larger than 40 degrees were only observed where the rocky terrain exists, which accounts for 3.51% of the site. In other words, naturally occurring slopes were in the range of 0 to 40 degrees except for rocky landscape in 1951. As seen in the slope map of the year 2009 in Figure 5.27 and Figure 5.28, most of the moderate slopes (30–40 degrees)

Figure 5.26 Perspective view of orthorectified Worldview-1 scene rendered on DEM highlighting the mine site (Emil, 2010)

Table 5.11 Collected data for the Ovacık coal mine (Emil, 2010)

Slope Range (degree)	Slope Type	Area in 1951		Area in 2009	
		m^2	Percentage	m^2	Percentage
0–30	Gentle slope	911132	85.64	890566	83.71
30–40	Moderate slope	115295	10.84	131273	12.34
40–50	Sloping	30047	2.82	31561	2.97
50–60	Steep slope	7113	0.67	7260	0.68
60–90	Very steep slope	265	0.02	3192	0.30
	Total	1063851	100	10663851	100

were accumulated in the borders of dump sites. This revealed that angle of repose for the dump material was between 30 and 40 degrees. The slope located in the north of the coal stock site was depicted as moderate slope both in 1951 and 2009. In addition to that, most of the very steep slopes, which are unstable (>60 degrees) are observed along the northern border of the open pit excavation.

The volume change polygons were overlaid with the slope map as seen in Figure 5.29. Also, areas of each slope range for each excavation and filling polygons were calculated and are given in Table 5.12. The fact that mining activities are major causes of steep unstable slopes is clearly observed from the difference between very steep slope percentages in disturbed and undisturbed land areas as 0.90% and 0.17%, respectively. Very steep slope presence in disturbed land is approximately five times more than in undisturbed lands caused by excavation and filling operations. Most of the very steep slopes throughout the mine site were located within the excavation polygon as also observed in the field works.

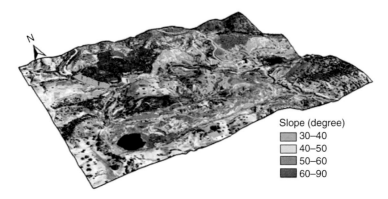

Figure 5.27 Perspective view of slope map overlaid with Worldwiew-1 satellite imagery (slopes lower than 30 degrees were not represented) (Emil, 2010) (See color plate section)

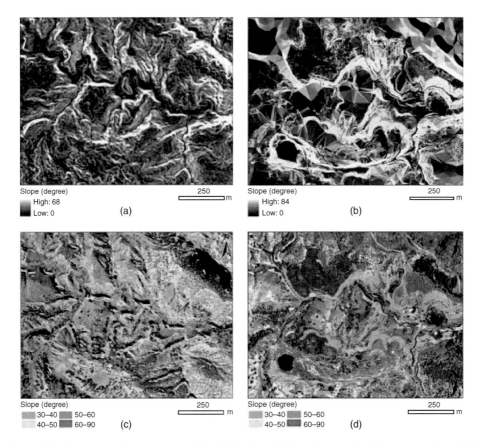

Figure 5.28 Slope maps of the mine site (a) in 1951, (b) in 2009 and reclassified slope maps (c) in 1951, (d) in 2009 overlaid with aerial orthophoto and Worldview-1 Satellite imagery respectively (in c and d, slopes lower than 30 degrees were not represented) (Emil, 2010) (See color plate section)

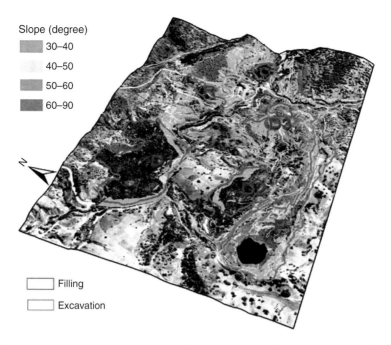

Figure 5.29 Perspective views of slope map overlaid with volume change polygons and Worldwiew-1 satellite imagery draped on DEM (slopes lower than 30 degrees were not represented) (Emil, 2010) (See color plate section)

Table 5.12 Area of land with respect to slope range and volume change polygons (Emil, 2010)

Slope Range (degree)	Disturbed Lands (m²)							Total Disturbed Lands (m²)	Undisturbed Lands (m²)	Total (m²)
	Excavation P	Filling								
		D1	D2	D3	C	R				
0–30	36286	59329	17412	9811	6591	4663		134093	756473	890566
30–40	16488	9487	4726	5680	57	1274		37712	93561	131273
40–50	5932	2034	418	448	28	89		8949	22611	31561
50–60	2238	478	47	32	5	0		2800	4460	7260
60–90	1658	13	0	1	0	0		1671	1520	3192
Total	62602	71341	22603	15972	6681	6026		185225	878626	1063851

REFERENCE

Blanco, J. L., Moreno, F. A., & Gonzalez, J. (2009) A Collection of outdoor robotic datasets with centimeter-accuracy ground truth. *Autonomous Robots*, 27 (4), 327–351.

Emil, M.K. (2010) *Land Degradation Assessment for an Abandoned Coal Mine with Geospatial Information Technologies*. M.Sc. Thesis, Middle East Technical University, Ankara, Turkey.

## Chapter 6

Remote sensing in mine reclamation

This chapter presents the role and significance of remote sensing in mine reclamation and closure practices. Following the definition of mine reclamation, the essence of mine reclamation for sustainable mining and its relevance to mine closure are stated. The main stages of mine reclamation from pre-mining planning to post-closure monitoring for surface mining are comprehensively discussed. Potential use of remote sensing as an alternative tool to conventional techniques at mine reclamation stages are explained in detail. Finally, two case studies about the use of remote sensing in surface coal mine reclamation are presented.

6.1 DEFINITION OF MINE RECLAMATION

All mining operations, whether they are surface or underground, are terminated after exploitation of mineral resources become economically and technically infeasible. Abandoning mines and mining areas without taking any measures causes various problems, such as, safety and health hazards to local people and habitat, degradation of land use, difficult rehabilitation conditions, downstream environmental impacts on land and water bodies, and public intolerance, which will eventually decrease the prestige of the mining business as well as diminishing mining industries' sustainability. In order to eliminate these negative impacts of abandoned mines, proper mine closure plans integrated with a mining project need to be prepared early in the process of mine development. Mine closure plans should be prepared to satisfy the requirements of regulations and expectations of local communities in order to:

- guarantee public health and safety,
- protect physical and chemical deterioration of natural resources,
- provide sustainable after-use of the mine site,
- minimize socio-economic impacts while maximizing socio-economic development.

Essential components of mine closure plans include mine closure design, closure operations integrated with ongoing mining activities, commitment to progressive site rehabilitation/reclamation, framework of short-term and long-term actions to be performed during mining activities, economic and social aspects for the local community, and financial planning. Commitment to effective mine closure necessitates detailed planning and implementation of mine reclamation, which is a process of returning an area, disturbed by mining activities, to its economical and ecological state so that it could serve for future generations without losing its capacity.

There are different terms that have been used for reclamation such as recultivation, rehabilitation and restoration, and reclamation. Although these terms are used interchangeably, there are important differences between them. Restoration brings about a situation where the land hosts conditions pertaining to those of the pre-excavation phase (Nicholson, 1988), foreseeing the dismantling of all constructions, with after-treatment of the land (Hartland-Swann, 1993). Rehabilitation, on the other hand, introduces a totally new aspect (Nicholson, 1988), to the land. However, the term of reclamation is often misinterpreted by individuals according to their viewpoints. Reclamation is believed to restore the resource components of the ecosystem to the pre-mining status of the environment, which cannot be achieved in many cases (Hossner, 1988). Reclamation is also considered as a revegetation. However, revegetation or planting is not equal to reclamation.

Munshower (1983) defined reclamation as follows:

> "*Reclamation includes all aspects of the environment; it is not restricted to soils and vegetation. Although the disturbed area cannot be returned to its exact premining condition, it can be rehabilitated. It can be returned to a useful function in the ecosystem of which it is a part. In all cases, however, the most economical means of attaining the reclamation goals is to develop a suitable reclamation plan prior to actual land disturbance.*"

As can be understood from this definition, reclamation is not considered as a simple post-mining operation, it is integrated with all stages of mining. It starts with pre-mining planning, continues through the exploitation stage, and ends with post-mining land use.

The main goal of reclamation is to return affected areas as near as possible to their economical and ecological value. It does not aim to return them to the original state (UNEP, 1983). In The Surface Mining Control and Reclamation Act (1977), which was passed in 1977 by the 95th Congress of the U.S.A. to regulate and enforce mine reclamation, the main elements and objectives of a successful reclamation are stated as follows (Hossner, 1988):

   i.   Revegetation efforts should be permanent, effective, and diverse,
  ii.   Post mine vegetative production should be considered equal to the approved standard when they are not less than 90% of the success standards,
 iii.   Restore land to a condition capable of supporting the uses which it was capable of supporting prior to any mining or higher or better uses,
  iv.   Prime farmlands require specified mine soil reconstruction techniques,
   v.   Minimize the impacts on the prevailing hydrologic balance on and off mine site,
  vi.   Restore approximate original topographical contours,
 vii.   General salvage of topsoil,
viii.   Identify and bury or treat potentially toxic or acid-forming materials,
  ix.   Minimize disturbances and adverse impacts on fish, wildlife, and related environmental values and achieve enhancement where possible.

6.2   NEED FOR MINE RECLAMATION

Mining inherently causes drastic disturbance of the mine environment and its surroundings. The level and extension of the mining induced impacts on the environment
depend upon the size of the mine site, method of the operation, ore type, and type of
impacts associated with the mining method. The mineral extraction techniques can be
classified into two main groups as surface and underground mining. Both methods can
have detrimental impacts on the Earth's surface, however, in surface mining a larger
surface area is disturbed due to the removal of the soil and overburden layers to expose
the deposit. The impacts of mining on the surface are not limited to the stripping of
strata above the mineral deposits and topographical change. Environmental and ecological impacts can be observed at each stage of mining. The relationship between
mining engineering and mine reclamation is presented in Figure 6.1.

Common impacts of mining outside of the mining site are classified into three
groups as water, air, and land. Pollution of water resources by heavy metals, chemicals,
and tailings disposals, alteration of surface runoff and underground water patterns are
common detrimental impacts of mining on water bodies close by the mining site. The
impacts on air varies from diesel emissions and dust into the atmosphere, to wind
erosion and air shocks. General consequences and impacts of mining on the ecology
of the mining site are as follows: (i) vegetation is destroyed, (ii) soil is disturbed, may

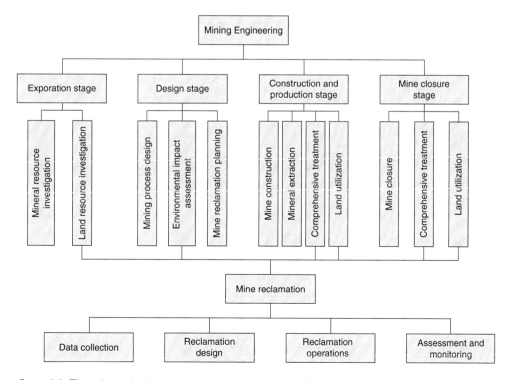

Figure 6.1 The relationship between mining and reclamation (Adapted from: Zhengfu and Dazhi, 1990)

be polluted, buried and mixed with other geological material, and surface materials remaining after mining are incapable of supporting plant life, (iii) wildlife habitats are greatly altered, and (iv) the resulting landscape is aesthetically displeasing. Besides the abovementioned ecological impacts, socio-economical and cultural changes could be developed due to mining activities. In order to eliminate these negative impacts of mining, mine closure and reclamation from planning to implementation is of vital importance to the mining industry, to communities, and to countries.

6.3 MINE RECLAMATION PLANNING

Mine reclamation is an integral component of mining and it aims to maximize production while maintaining environmental, socio-economical, and ecological quality of the mine environment with the constraints imposed by the regulatory authorities and the needs to satisfy the economical objectives of the mining company. Since mine reclamation is not a simple process it requires careful planning for each stage associated with the reclamation practices.

Planning and construction of mine reclamation is fundamental to nearly all aspects of the reclamation operation. A mine reclamation plan, which defines the measures to be taken after mine closure, is essential for sustaining minimized cost of reclamation and minimized production operations interference. Reclamation plans address direct and indirect impacts of mine closure on environmental, ecological, and socio-economical state of the community in local, regional, and global extent. Reclamation planning involves the consideration of:

1 Land use prior to mining and natural and cultural factors affecting the land use,
2 Land capability prior to mining, to support a variety of uses,
3 The alternatives for land use after completion of mining and reclamation,
4 A detailed description of how the proposed post-mining land use is to be achieved,
5 Methods for ensuring that mining and reclamation operations are in line with declared land use plans and programs,
6 Mining methods used to ensure that the reclamation plan is compatible with local physical, environmental, and climatic conditions,
7 The economical and technical feasibility of the reclamation plans, and
8 The current mining and regulatory conditions.

The necessary planning steps are:

i.   make an inventory of pre-mining conditions,
ii. evaluate and decide on the post-mining requirements of the area consistent with the needs and desires of the local communities,
iii. analyze alternative mining and reclamation schemes to determine best alternatives meeting overall objectives,
iv. review reclamation experience at neighboring mines or at mines operating in similar conditions to determine successful reclamation practices and parameters, and
v.   develop an acceptable mining, reclamation and land use scheme to be integrated with existing technical, social, political, and economical conditions.

The extent and content of the reclamation plans vary according to the status of mining and the stage of the mine's life. Reclamation planning is prepared and performed at early, routine, and post-mining stages of mining. Remote sensing applications are widely used in early and post-mining stages for monitoring the performance of revegetation activities as well as the topographical changes. It is also possible to use remote sensing in reclamation for routine mining activities. The use of remote sensing in the routine mining stage is mostly monitoring the revegetation. However, remote sensing has a variety of applications in reclamation planning for the pre-mining stage. Hence, the following subsections provide possible uses of remote sensing in pre-mining reclamation planning.

6.3.1 Reclamation planning in pre-mining stage

Since mining is a business which uses a large areal extent of land, land use planning of the mine area is of vital importance for reclamation planning and practices in order to define the characteristics of the area before the mining activities initiate. Table 6.1 shows the percentage of land use by mining activities depending on the type of the operation and type of commodity. Effective land use planning forms the basis of sustainability. If there is no land use plan existing for the site, then it needs to be prepared. In this process, the land use classification standards of the country, providing planners with a consistent methodology for classifying land uses based on their characteristics, have to be followed. Therefore, remote sensing provides rapid development of existing land use maps for a given potential mine site.

6.3.2 Baseline study

Mine reclamation planning requires having sufficient environmental baseline information to set the basic reclamation targets based on the original status of the environment and to support the successful implementation of reclamation. Collecting the baseline information is called a **baseline study**. Baseline study can also be defined as a characterization of the mine area and its surroundings quantitatively and qualitatively

Table 6.1 Land use by mining according to the commodity and operation (Frimpong, 2003)

Percentage of Land Use by Mining Allocated by Commodity		Percentage of Land Use by Mining Allocated by Mining Function	
Bituminous Coal	48%	Mined Area (Excavated)	69%
Sand and Gravel	17%	Waste Area from Surface Mining	16%
Stone	13%	Processing Waste Area	10%
Phosphate Rock	5%	Waste Area from Underground	3%
Iron Ore	2%	Subsided Area	2%
Clays	4%		
Copper	3%		
All other minerals	8%		

to establish the baseline level of potential for contaminants in land, water, and air quality within the concerned site, and to assess the extent of site deterioration.

The main stage of the baseline study involves collection of data and information about each land use class and the mine environment for planning reclamation.

Some of the data can be available from official resources. The rest must be collected by the company during exploration. If the data are not obtained until the mining operations initiate then historical data may be required. In this sense, remote sensing plays an important role especially in collecting historical data about a specific site. Conventional techniques such as topographical measurements and surveying of the pre-mining status of the mine environment may not present the original form of the environment. In this case, archived satellite data can be used to obtain data about the pre-mining conditions of the mine environment for the baseline study.

Any collected data should be stored in a database including spatial and temporal nature of the data. GIS is also utilized to store and retrieve the obtained spatial data. Especially for large surface areas or areas of difficult access, remote sensing can effectively be utilized for baseline study. Different spectral bands of satellite data can be used for the collection and analyses of different features in the baseline study. The main components of baseline study data and the types of remote sensing data are given in Table 6.2.

Table 6.2 Essential features of a baseline study

Essential Baseline Data Groups	Type of Data	Potential Remote Sensing Data and Analysis
Climate	Precipitation, elevation, temperature, atmospheric gas concentrations, snow, etc.	High resolution satellite images, synthetic aperture radar (SAR), lidar
Biological	Plant population, vegetation health, plant disease, wild life, etc.	Aerial photography, satellite images, lidar and synthetic aperture radar (SAR)
Geological	Soil, rock strata, structural geology, discontinuities, etc.	Aerial photography, hyper spectral remote sensing data, synthetic aperture radar data (SAR)
Hydrological	Ground water resources, quality and quantity of water resources, physical and chemical properties of water, etc.	Aerial photography, satellite images, radar data
Soil and overburden material	Properties of soil: pH, conductivity, sodium adsorption ratio (SAR), cation exchange capacity (CEC), metallic properties, texture, etc.	Aerial photography, satellite images, gamma-ray spectroscopy, radar systems
Topographical	Elevation, slope, aspect, etc.	Orthorectified satellite images, aerial photography, digital contour maps
Cultural, historical and socio-economical	Population, GDP per capita, etc.	High resolution satellite images (e.g. archeological findings)

The essential data for the baseline study are listed as follows.

- *Topographical:* Data about elevation, slope, aspect, morphological features in the mine site are required for appropriate mine layout design like selecting proper dump site, tailing dam, etc. The contour maps of the area in various scales (1:25,000, 1:10,000, 1:5,000, 1:1,000) should be obtained. Based on the obtained digital contour maps digital elevation models (DEM), slope and aspect maps can be obtained. Stereo satellite images and aerial photos are effectively used in order to generate DEM for the topography.
- *Climate:* Data about the climate are required for (i) identifying hydrologic conditions, precipitation, (ii) analyzing stability of dumps and stock piles by analyzing the amount of seasonal distribution of rainfall, periods, duration, and the amount of maximum precipitation for flood forecasting, (iii) controlling erosion by collecting annual precipitation, temperature, wind speed, and direction data, (iv) revegetating the land based on the maximum, minimum, and mean temperatures, for severe winter conditions, duration of frost-free weather, and (v) identifying atmospheric concentrations of carbon dioxide, carbon monoxide, and ozone.
- *Biological:* Data about the biodiversity and biological characteristics of the area is required for revegetation of the land and restoration of the fauna and the wildlife of the mine environment for reclamation purposes. The essential biological data for the baseline studies can vary from the plant population to the types of wild animals. Plant population should be predicted for each plant type. The prediction is done by counting different species of plants, their relative abundance, total plant density, and productivity by selecting usually an approximate area of $1\,m^2$ from $100\,m$ grids. Predicting plant population by this method requires costly, time consuming, and labor intensive field work which may become infeasible in practice. In this sense, remote sensing techniques play a vital role especially for inaccessible large surface areas to detect and monitor plant estimates, spread of plant disease, selective deforestation, plant species abundance profile, and other detailed vegetation data. Lidar and Synthetic Aperture Radar (SAR) can be utilized for plant population prediction. Land use/cover maps can be obtained from the remote sensing images or aerial photos. Land cover or land use maps include the total land area, arable and permanent cropland, permanent pasture, forest and woodland, residential areas, etc. Animal population should also be determined by counting or by capturing the known or endemic mammal, bird and fish species. The proportion of tagged animals recaptured is a measure of population. Also farm animals as stock of cattle, sheep, goats, pigs, equines, buffalo, and camels should be counted and recorded within the biological data. If the land is used for agriculture, the yields of various crops/acre should be recorded. If the land is used for grazing, the number of acres required/animal should be recorded. If the land is productive forest, productivity in m^3 of timber should be recorded.
- *Geological:* The geological mapping and succession of material from surface to the top of the ore to be mined is required. The geological data obtained for the baseline study should include: (i) soil, topsoil, subsoil, rooting zones with the contacts between them, (ii) material below the subsoil which requires special treatment, such as sulphide-containing rocks which should be stripped separately and buried

Table 6.3 Typical swell factors for overburden of coal seams (Muir, 1979)

Material	Swell Factor (%)
Compact dry clay	33
Mixed, wet or dry clay, gravel	40
Dry loam (earth)	15–35
Moist loam (earth)	20–30
Wet loam (earth)	20–35
Mixed earth, sand, gravel	18
Mixed earth, rock	30
Dry gravel	14
Wet gravel	18
Limestone	65
Well blasted rock	50
Sandstone	50
Dry, moist or wet sand	14
Shale	33

in the dump to avoid acid mine drainage, and clay which can cause instability problems, (iii) variability of geological conditions which determines the density of drilling and a geological map of the site, (iv) structural geology of the area including faults, folds, joints, bedding planes, and other discontinuities, (v) structural feature maps to be used for evaluating the stability of ground and working faces, and (vi) the prediction of amount of swell to estimate the degree of refilling the void and capacity of dump sites. The importance of swell factor is highly recognized in dump site selection and design. Table 6.3 presents the differentiation of swell factors for different structures.

Remote sensing applications in geology provide a valuable tool to obtain geological data for the baseline study. Using multispectral satellite data, information on lithology, rock composition or rock alteration, can be provided. Radar data provides an expression of surface topography and roughness, and thus is extremely valuable, especially when integrated with other data sources in providing details of relief or physiography. The abilities of remote sensing are not limited to surface features, it can also be used to investigate the geological structures beneath the surface using the condition of vegetation growing at the surface, which is called geobotany. The underlying principle is that the mineral and sedimentary constituents of the bedrock may control or influence the condition of vegetation growing above. Current technological developments in the remote sensing allow one to achieve various projects using remotely sensed data including surficial deposit/bedrock mapping, lithological mapping, structural mapping, sand and gravel (aggregate) exploration/exploitation, mineral exploration, hydrocarbon exploration, environmental geology, geobotany, baseline infrastructure, sedimentation mapping and monitoring, event mapping and monitoring, geo-hazard mapping, and planetary mapping.

- *Hydrological:* Usually water resources are more sensitive to disturbance than the land is. Surface water resources such as river, stream, lake, sea, groundwater resources, and level of water table, are mainly affected by mining activities. For

Table 6.4 A typical database obtained from water quality analysis

Acidity	Conductivity
Alkalinity	Dissolved solids
Ammonia_N	Hardness
Arsenic	Metals (Al, Cu, Cr, Cd, Pb,
Biochemical Oxygen Demand	Mn, P, K, Ni, Zn, Na,
Boron	Fe, Hg, Ba, Ca, Ag, Se)
Chloride	pH
Chemical Oxygen Demand	Sulfate
Coliforms	Suspended solids
	Turbidity

the baseline study the typical data need to be obtained mainly include physical and chemical properties, results of water quality analyses, base maps of surface drainage systems, present use of water related to water rights, and discharge of streams. Table 6.4 presents the typical data required to be obtained for water quality analysis.

Remote sensing through image interpretation techniques, especially radar technology with its active sensing capabilities, offers an effective tool to capture the spatial distribution and dynamics of water resources, which are frequently not accessible using traditional ground surveys. Optical remote sensing data can also be utilized when it is not constrained with cloud conditions. The remotely sensed measurements such as temperature and backscatter of active microwaves from saturated zones can be of use for groundwater monitoring studies. A more sophisticated approach using thermal and multispectral imagery was adopted by Thunnissen and Nieuwenhuis (1989) to determine the effect of groundwater extraction on crop evaporation. The drainage networks for the site can also be obtained from DEM extracted from stereo image pairs, LIDAR data or interferograms.

- *Soil and overburden material:* The properties of soil that must be established depend on the existing and proposed use. Soil conditions before mining were used as the standard or reference condition for the reclaimed soils in the determination of a soil quality. If farm or agricultural land is involved, detailed data about the soil and overburden is necessary. The general definition and classification of soil and subsoil, which form the rooting zone of plants are presented in Figure 6.2 and the general properties that must be analyzed are listed in Table 6.5.

O-Horizon: The uppermost horizon that is mostly made up of decaying organic matter, also called humus.

A-Horizon: The upper layer beneath the O horizon, it is called the top soil. The organic matter is most abundant in it and the leaching of soluble salts is the greatest. Seeds sprout and plant roots grow in A horizon layer.

E-Horizon: The layer beneath the A horizon. The soil has lighter color. It is mostly made up of sand and silt. A leaching process takes place in the E horizon which allows water to drip through the soil to carry away most of the minerals and clay originally present. Also called eluviation layer.

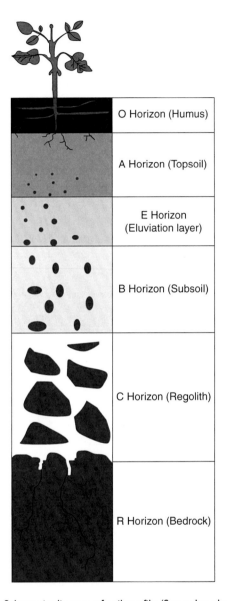

Figure 6.2 Schematic diagram of soil profile (See color plate section)

B-Horizon: The next deeper layer, the sub soil. It contains more clay, iron and alumina than A and C zones. Sometimes called illuviation layer.

C-Horizon: The deepest layer. It is composed of loose material or weathered rock and is relatively unaffected by biological action. Plants and roots do not grow in this layer and also very little humus is found in this layer. This layer is also known as regolith.

Table 6.5 Typical data obtained from soil quality analyses

pH	Measure of acidity or alkalinity. Between 7.2 and 8.2 is optimal for growth. No growth above 8.2 and below 3.5. Between 7.2 and 3.5 the number of species is limited.
Conductivity	Measure of salinity. Electrical conductivity related to the concentration of soluble salts in soil solution. Increased conductivity indicates reduced growth capacity. Below 4 slight effect, 4 and 8 moderate effect and 8–16 severe effect.
Sodium Adsorption Ratio (SAR)	Measure of relative abundance of sodium ions in ionic exchange reactions in soil. Below 4 shows a soil favorable to growth, above 10–15 is unfavorable.
Cation Exchange Capacity (CEC)	Measure of soil capacity to absorb and hold cations. It is an important indicator of nutrient status as exchangeable cations are the most important source of plant nutrients.
Metallic Properties	The pyritic material in the overburden is the main source of acidity and should be buried in the dump. The base metals (i.e. Non-ferrous metals excluding precious metals, Cu, Al, Pb, Ni, Sn and Zn) may also give acid toxicity in soil heaps and mine water. Rare metals (boron, selenium, molybdenum, cadmium and mercury) are also potentially toxic.
Texture	For clay, silt, sand texture and color are also important for absorption of heat.

R-Horizon: If there is a solid rock layer underneath all of the horizons except R, then it is part of the R horizon. This layer is also called bedrock.

General soil properties and typical data that should be obtained for soil quality analyses for productivity in revegetation are listed in Table 6.5.

Remote sensing techniques can contribute important supporting information for digital soil mapping and provide effective measurement and monitoring of soil especially when physical measurements cannot be made. Passive, airborne or satellite remote sensing systems have widely been used for terrestrial monitoring and for obtaining surface vegetation data. Active remote sensing systems can deliver the digital terrain data which possibly is the most important contribution to soil monitoring. Currently, the bias and precision of available radar data sets is in the order of meters (better data are collected but access is restricted for military reasons) (Wood *et al.*, 2004). Also thermal systems, which have also a potential for soil monitoring through the thermal capacities of soil, and gamma-ray spectroscopy, which can gather geological information, are other remote sensing systems to be utilized for soil monitoring.

- *Cultural, historical, and socio-economical:* Cultural, historical, socio-economical, and demographical data including total population, current and projected population density, distribution of the population to urban and rural areas, etc. should be collected within the region. Natural and cultural factors to be addressed in reclamation planning are summarized in Tables 6.6 and 6.7, respectively. The relative importance of natural and cultural factors in land use suitability is given in Table 6.8 and Table 6.9, respectively. Also, GDP per capita, GDP from agriculture, from industry and from services, and GDP annual growth rate should be obtained.

Table 6.6 Natural factors related to mine reclamation (Frimpong, 2003)

I. NATURAL FACTORS		
A. Topography	1. Relief 2. Slope	
B. Climate	1. Precipitation 2. Wind-airflow pattern, intensity 3. Humidity 4. Temperature 5. Climate type 6. Growing Season 7. Microclimatic characteristics	
C. Altitude		
D. Exposure		

E. Hydrology	1. Surface hydrology
	  a) Watershed considerations
	  b) Flood plain delineations
	  c) Surface drainage patterns
	  d) Amount and quality of runoff
	2. Ground water hydrology
	  a) Ground water table
	  b) Aquifers
	  c) Amount and quality of ground water flows
	  d) Recharge potential	

F. Geology	1. Stratigraphy 2. Structure 3. Geomorphology 4. Chemical nature of overburden 5. Coal characterization	

G. Soils	1. Agricultural characteristics
	  a) Texture
	  b) Structure
	  c) Organic matter content
	  d) Moisture content
	  e) Permeability
	  f) pH
	  g) Depth of bedrock
	  h) Color
	2. Engineering Characteristics
	  a) Shrink-swell potential
	  b) Wetness
	  c) Depth of bedrock
	  d) Erodibility
	  e) Slope
	  f) Bearing capacity
	  g) Organic layers	

H. Terrestrial Ecology	1. Natural vegetation, characterization, identification of survival needs 2. Crops 3. Game animals 4. Resident and migrant birds 5. Rare and endangered species	
I. Aquatic Ecology	1. Aquatic animals-fish, water birds, resident, and migratory 2. Aquatic plants 3. Characterization, use, and survival needs of aquatic life system	

Table 6.7 Cultural factors related to mine reclamation (Frimpong, 2003)

2. CULTURAL FACTORS	
A. Location	
B. Accessibility	1. Travel distance
	2. Travel time
	3. Transportation networks
C. Size and shape of the site	
D. Surrounding land use	1. Current
	2. Historical
	3. Land-use plans
	4. Zoning ordinates
E. Land ownership	1. Public
	2. Industry
	3. Private
F. Type, intensity and value of use	1. Agriculture
	2. Forestry
	3. Recreational
	4. Residential
	5. Commercial
	6. Industrial
	7. Institutional
	8. Transportation/Utilities
	9. Water
G. Population characteristics	1. Population
	2. Population shifts
	3. Density
	4. Age distribution
	5. Number of households
	6. Household size
	7. Average income
	8. Employment
	9. Educational levels

6.4   MINE RECLAMATION OPERATIONS

Mine reclamation is an integral part of all mining activities and is especially important in surface coal operations which are of large areal extent. The unit operations and major steps in reclamation for strip mining are summarized as follows:

   i.   Soil, overburden, hydrology, vegetation, and wildlife inventory studies,
  ii.   Mining permit application preparation,
 iii.   Topsoil removal and stock piling it to a suitable site,
  iv.   Subsoil stripping,
   v.   Overburden stripping and placement in the dump site,
  vi.   Mining the ore (usually coal in strip mining),
 vii.   Spreading the stacked overburden material,
viii.   Grading of overburden and leveling the spoil piles,

ix. Topsoil replacement, and
x. Revegetation.

The abovementioned stages of reclamation should be conducted sequentially and in coordination with each other. They are initiated prior to mining and continue after mine closure for monitoring and controlling the success of reclamation.

Surface mining reclamation operations start with the removal and stockpiling of the topsoil, and where necessary subsoil, on the surface of the mine site. Topsoil and subsoil layers have to be removed and stockpiled separately. In strip mining, normal practice is to stockpile the soils from the initial ground clearing into long term storage and then to take the soils removed as part of the ongoing operations and place them directly into the areas to be reclaimed. After removing the soil layers and stockpiling it in a suitable site, overburden material is stripped and put in a dump site so that the coal or ore layer is exposed to the surface and got ready for excavation.

Grading the piles of the soil and revegetation of the area is performed by planting and other work necessary to restore an area of land. The type of vegetation depends on the original vegetation of the land before mining activities are carried out and agreed post-mine use of the land. The restoration of the area can be done for different purposes such as, forest growth, grazing, developing agricultural areas, recreational or wild life purpose or some other useful purpose of equal or greater value of land. Mine reclamation requires a proficiency in various disciplines.

In the case of open pit mining (primarily referring to metalliferous ore), there are stages differing from the steps in reclamation for strip mining due to the nature of the mining method. Open pit mining requires waste dumps to dump the stripped overburden from the pit and tailing dams to store the tailings and chemical wastes from the processing plants. Figure 6.3 presents a tailing dam of a copper mine. Any failure

Figure 6.3 Tailing dam of a copper mine (See color plate section)

in waste dumps slopes and tailing dams may cause deferred production, machine and equipment loss, injuries and fatalities, and environmental problems.

Therefore, particular attention should be paid in designing and constructing solid waste dumps and tailing dams to ensure their physical and chemical stability. Erosion and sediment transportation should also be prevented.

Due to the nature of surface coal mines, they are the ones which have the highest land disturbances. Especially reclamation works for surface coal mines have their own characteristics as they introduce the highest land disturbance due to the layered nature of the coal bed. The sequential reclamation activities for surface coal mining are given as follows:

I.   **During Site Preparation**

1   Install control measures (diversion, sediment traps and basins, etc.)
2   Clear and grub market lumber if possible, stockpile brush chips for mulch
3   Stabilize areas around temporary facilities such as maintenance yard, power station, and supply.

II.   **During overburden removal**

1   Divert water away from mining areas
2   Remove topsoil and store it if possible and/or necessary
3   Selectively mine and place overburden strata if possible and/or necessary

III.   **During coal removal**

1   Remove all coal insofar as possible
2   For the purpose of controlling post-mining groundwater flows, break, or conversely prevent damage to, the strata immediately after coal removal

IV.   **Immediately after coal removal**

1   Seal the highwall if necessary
2   Seal the longwall if necessary
3   Backfill, bury toxic materials and boulders, dispose of waste, and ensure compaction

V.   **Shortly after the coal removal**

1   Rough grade and contour, taking these factors into consideration
    a)   Time of grading-specific time limit, tied to advance of mining, seasonal considerations
    b)   Slope steepness
    c)   Length of uninterrupted slope
    d)   Compaction
    e)   Reconstruction of underground and surface drainage patterns
2   If necessary, make mine spoil amendments (root zone), taking these factors into considerations
    a)   Type of amendment-fertilizers, limestone, fly ash, sewage sludge, or others
    b)   Depth of application

c)   Top layer considerations-temperature (color), water retention (size, consistency organics), mulching and tacking

VI.   **Immediately prior to first planting season**

1   Fine grade and spread topsoil, taking seasonal fluctuations into consideration
2   If necessary, manipulate the soil mechanically-ripping, furrowing, deep-chiseling or harrowing, or constructing dozer basins
3   Mulch and tack

VII.   **During the first planting season**

Seed and revegetate, considering time and methods of seeding, choice of grasses and legumes

VIII.   **At regular, frequent intervals**

Monitor and control slope stability, water quality, both chemical (i.e. pH) and physical (sediment), vegetation growth, etc.

6.4.1   Soil and subsoil removal and replacement

Soil as a vital resource for sustaining human lives should be removed and kept prior to stripping overburden material. In soil removal, topsoil and subsoil layers should be removed selectively and stored separately. The waste dumps should be leveled or contoured for stability. After accessing the ore and excavating it, the stored soil layers should be placed back and spread for revegetation.

Soil removal is performed using various types of equipment and machinery such as scrapers, front-end loaders and dumpers, bulldozers feeding to front-end loaders and dumpers. For leveling the dump and spreading the subsoil, bulldozers and scrapers are used in combination, for spreading the topsoil, scrapers, graders, and draglines alone or in combination can be utilized.

Since cable shovels are heavy duty machines with compressive action in scraping off a vertical side of face, they are less adapted to selective mining. If overburden requires blasting then selective mining also requires selective blasting. To pick up spoil selectively is half of the problem. To dispose of it selectively in the dump is equally difficult, because this requires arrangement of dumping areas in the lower part of the dump for a specific material from the working face.

6.4.2   Soil and subsoil storage and protection

The soil picked up during the development of the mine should be stored for when the time comes for resoiling of the waste. In a mining area of limited extent and in a short term mining operation all the topsoil can be stripped and stored in one operation. In long term workings, which occupy an extensive land area, the topsoil is removed ahead of the advancing face. When the face is far enough advanced to give room in the pit for the waste, the top soil can be transported directly to its point of use for replacement.

The stored soil pending its ultimate replacement must be protected against erosion and pollution by grading the dump with gentle slopes and vegetated to give quick plant cover. The removal of top soil can cause air and water pollution. Selected overburden material (i.e. material below the subsoil) may be used instead of top soil if it is suitable for vegetation.

6.4.3   Overburden removal and dumping

Overburden is defined as a material layer above the ore body that must be removed or stripped in order to expose the ore to the surface. The stripped overburden is relocated to the dump sites. The location of dump sites should be decided by considering stability, amenity, and transportation cost of overburden. In the case of strip mining, dump sites are mined out areas just behind the operating highwalls. In open pit mining, dump sites are distant to the pit. If the dump will finally be used to fill the final void it may be built as close to that area as possible to reduce the transportation cost. If the dump is a permanent feature, it may be located where it will best blend with the landscape or fill depressions with minimum drainage disturbance. The dump site should provide a foundation capable of giving firm support and stability and should not interfere with natural drainage or in such a way that it allows flow of surface water into streams, surface run-off or springs.

The key steps in the design of waste dumps are summarized as follows:

 i.   Determination of alternative sites,
 ii. Comparison of the area, height, slope, and economics of construction for alternative disposal sites,
 iii. Analysis of geology related to dump site and overburden characteristics,
 iv. Analysis of surface drainage network,
 v.   Analysis of internal and external drainage requirements,
 vi. Identification of overburden removal procedures, and
 vii. Analysis of needs for dump site construction.

Stability of the waste dumps mainly depends on the nature of the ground on which the dump site is built, therefore, the ground characteristics should be established during exploration by identifying the bearing capacity and shear strengths of the layers in the dump site. It is known that the bearing capacity affects the rate of dumping and final height. The dip and orientation of the overburden layers also affect the stability. It is observed that building up the dump in successive horizontal layers permits higher dumps.

The other important factor affecting stability is the internal water conditions. If water penetrates and accumulates in the dump, it will reduce the shear strength of the material, cause flow or sweep from the face at the toe and cause instabilities and erosion.

Characteristics of the spoil material with or without compaction affect stability. Compaction reduces pore spaces and hence reduces the capacity to build up water pressure. It also increases the proportion of rainfall that flows off the surface of the dump and the amount of material to be dumped. Therefore, density, porosity, swell factor, internal angle of friction, and frictional strength need to be determined for achieving the optimum shape, height, and slope angle design for the spoil pile or waste dumps.

In addition to considering the factors affecting spoil pile stability, there may be precautions that can be taken against failures in dumps, which are listed as follows:

- Select sites with the most moderately sloping and naturally stable grounds.
- If it gives additional stability place spoil above a natural terrace bench or berm.
- Build key cuts (excavations or key cuts) or rock toe buttresses, if the spoil is dumped on steep slopes (>36%).
- Prevent water intake in the dump areas, unless natural drains are constructed.
- Dump the spoil in a controlled manner to avoid mass movements.
- Cover or grade the dump to aid surface and subsurface drainage compatible with natural drainage.
- If the spoil is dumped to natural valleys, the site must be near the ridge top of the valley to increase stability and reduce drainage above the dump.

In order to avoid pollution and minimize the waste from the mining site, revegetation practices should be started as early as possible in dump sites. The slopes at the vegetated dump sites should be optimized to avoid any erosion and to increase productivity. Water from workings and run-off in settling ponds needs to be collected in settling ponds, to allow solids to separate before returning the water to natural water resources.

In order to avoid acid mine drainage, contact of water with sulfur-bearing spoil should be prevented. If water is polluted despite precautions, then it must be treated by neutralization, precipitation or biological correction. Figure 6.4 illustrates an example of acid mine drainage. Water treatment by neutralization is achieved by adding lime, limestone, dolomite or magnesite to the water to neutralize the pH of water. Treatment can also be done by precipitation using lime. Biological correction requires encouraging growth of algae to trap heavy metals.

6.4.4 Revegetation of dump sites

Revegetation is the most visible and thus the most important phase of the reclamation process. Regardless of the effort and investment which have been put into grading, drainage, and topsoiling, reclamation cannot be considered successful until there is full growth of plants, shrubs or trees on the disturbed land. The leveled and graded dump sites are revegetated after exploiting ore deposits. Figure 6.5 illustrates an example of revegetation practices during planting of trees.

The main objectives of revegetation are to stabilize the dump and prevent erosion and to improve the appearance of the site. For revegetation, the selection of the plant species has a significant role for effective reclamation practices. The appropriate plant species or trees will not only provide food for wildlife but also stabilize the soil by replacing nitrogen. For example, in soil like temporary dumps, which can be used to backfill the final void, grasses are effective, however on the site slopes deeper rooting plants should be preferred to increase stability. Naturally occurring grasses, shrubs, and trees, etc. will generally grow more quickly and successfully than species imported

Figure 6.4 Acid mine drainage in an abandoned coal mine (See color plate section)

Figure 6.5 Revegetation practices during planting of trees

from other areas. The existing terrain, as well as the climate of the area, plays a big part in the proper selection of vegetation for reclamation.

The major steps followed in revegetation are given in Figure 6.6. As can be seen from Figure 6.6, the revegetation process is initiated by obtaining the soil which was removed and stocked prior to overburden stripping operations. The stored subsoil and

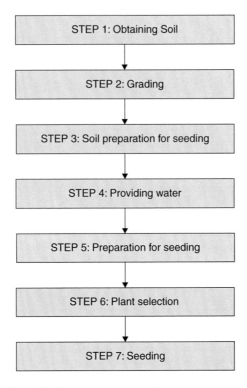

Figure 6.6 Revegetation stages in mine reclamation

topsoil is transported to the dump sites and put back as layers on top of the overburden material successively. This process constitutes the major portion of the reclamation cost and is dependent upon the type and location of the stock piles where the soil is maintained. After placing the soil back, the next step is to contour the spoil piles for meeting the profiles required for the particular location. The contouring process is predominantly carried out using dozers and sometimes small capacity draglines. The major problem faced in revegetation is to revitalize the topsoil to the level where it can sustain useful plant growth. There is no universally accepted method to accomplish revitalization. Each situation will have a different set of variables such as local topsoil quality, quality of the underlying strata, and climate.

The second stage of the revegetation is grading, amount of which varies considerably depending on the type of revegetation (Figure 6.6). Grading guidelines dictates that little grading is required for forestry, more grading is required for grazing land, and extensive grading is required for agricultural applications. If the area is reclaimed by recreational activities or by real estate then there is no need for grading and a lake can be left in the final cut (Figure 6.7). An optimum grading should be achieved due to the undesired effects of too much grading, which causes compacting of the soil. Insufficient grading causes puddling, in which plants rot in wet weather, and baked mud flats in dry weather. Contouring associated with grading is usually tied into the surrounding

Figure 6.7 Lake formed in the pit of an abandoned mine

area for (i) providing good slope stability with good drainage, (ii) controlling surface runoff to hold water where necessary, and (iii) recognizing the future land use.

The next stage is soil preparation for seeding, and requires removing of stones and rocks larger than a certain size using a mechanical stone picker (Figure 6.6). In the first 60 cm of the soil layer, the amount of rock particles greater than 15 cm should be less than 20% of the volume of the associated layer and in the second 60 cm of the soil layer, the volume of rock particles greater than 25 cm should be less than 50% of the associated soil layer. Also, the rock volume should not exceed 20% of the total volume.

Providing water is the next stage of the revegetation process (Figure 6.6). Water is not only necessary for seed germination but also as a leaching agent. As it percolates into the soil, water reduces the acidity of the soil and leaches out the high levels of soluble salts at the soil surface. Water also serves to reduce the surface amount of trace metals such as manganese, copper, and zinc which may be toxic at high concentrations. After the preconditioning and upgrading of the soil surface is completed, planting begins (Figure 6.6). The first step is plant selection. The plants must be matched according to climate, length of growing season, amount of rainfall, quality of soil, and land contour and eventual use of the land. High quality soil can easily turn into crops or grazing, on the other hand lower quality soil requires plants or trees that are not heavy nutrient consumers and which can thrive in slightly acidic soil.

Providing water is followed by the stage in which the soil is prepared for seeding by placing mulch, a fibrous material, on top of the soil. The most common mulch materials are straw, hay, and wood fiber. Mulch improves moisture availability and consequently enhances germination and seedling growth. Other effects are (i) to

Figure 6.8 Revegetated dump site of an abandoned mine site

moderate soil temperature, (ii) to provide a shading effect for the tender germinating seeds, (iii) to provide a stable seed bed by reducing the effects of wind and water, (iv) to reduce erosion particularly raindrop impact, and (v) to encourage rapid moisture infiltration and reduced evaporation.

After preparation for seeding, the plant type which is compatible to the soil and natural vegetation, is selected (Figure 6.6). Generally, soil higher in phosphorous is appropriate for agricultural use, while those lower in phosphorous are more suited to tree growth. There are no set rules for plant selection, however, the plant must match the characteristics of the site. Trees are adaptable to a wider range of soil conditions than plants. European alder and black locust are popular trees adding nitrogen to the soil. Pines are popular in sandy soils. Legumes add nitrogen to the soil. Under better soil conditions cypress, poplar, walnut, oak, maple, chestnut, and locust can be planted for revegetation. Figure 6.8 illustrates a dump site revegetated by locust trees in an abandoned mine. There are several seeding methods such as hand seeders (mounted or shoulder), hydroseeding (fertilizer + seeds + mulch comb → slurry), spread under pressure from a tank or truck (needs road access), and aerial seeding in which a plane or helicopter is used. Tree planting is done using nursery seedlings planted by hand or machine.

6.5   RECLAMATION CONTROL AND MONITORING

Sustainable mining requires continuous monitoring of the changes in the mine environment to identify the long-term impacts of mining on the environment and land cover to provide essential reclamation practice and to check the effectiveness of reclamation.

As the last and the one of the most important stages of mine reclamation, reclamation control and monitoring of the reclaimed area is performed on a regular basis to ensure that the environment is restored to conditions which are set in the closure and reclamation plan. The main objective of control and monitoring is to check whether reclamation objectives are achieved and commitments are met or not. The essential processes in this stage are (i) control and monitoring of surface and underground water, (ii) progress and health of revegetation, (iii) topographical change and stability, and (iv) wildlife and habitat, which are essential to observe the success of reclamation using objective and accurate data. Usually land reclaimed for agriculture and grazing requires at least five years of monitoring. Restoration for forestry may require longer monitoring times. Considering the frequency of the monitoring and the scale of the area to be monitored conventional techniques may become costly and impractical. Conventional techniques in mine land monitoring such as topographic measurements are time consuming and labor intensive, therefore are not efficiently used for a large-scale surface mine area (Anderson *et al.*, 1977). In this sense, remote sensing provides a valuable tool to obtain rigorous data and hence diminishes the essence of time-consuming and costly field measurements. The use of remotely sensed satellite data has been widely applied to provide a cost-effective means of supplying land cover mapping over large geographic regions in a quicker time. Therefore, monitoring vegetation growth and health during reclamation works greatly benefits from remote sensing.

The main stages of control and monitoring using remotely sensed data includes: (i) acquiring multi-temporal satellite data, (ii) preprocessing of the obtained data, (iii) comparing the images and detecting the change in the area using remote sensing image interpretation techniques and change detection algorithms (Lu *et al.*, 2005). The type of sensors and the characteristics of the data may vary depending on the size and nature of the area and the variability of the features to be monitored. Different types of sensors and the properties of their data are given in Appendix A. In reclamation monitoring, detecting topographical stability through temporal DEM analysis and assessing the changes in land use and land cover are the two widely used tasks using remote sensing.

Change detection basically aims to identify, quantify, and analyze the spatial response of landscape due to the mining activities. The fundamentals and principles of change detection algorithms are explained in detail in Chapter 3.

The selection of an appropriate change detection algorithm is important (Jensen *et al.*, 1993; Dobson *et al.*, 1995). First it has a direct impact on the type of image classification to be performed (if any) (Lu and Weng, 2007). Second, it dictates whether important temporal information can be extracted from the imagery. Kaufmann and Seto (2001) proposed that the change detection method depends on the landscape of the study area, the kind of change that occurred in the landscape and the temporal and spatial characteristics of the data available.

In order to monitor the vegetation cover or the progress of revegetation, indices, i.e. NDVI, derived from the spectral bands of satellite images can be utilized effectively. Different vegetation indices, their mathematical representation and principles are comprehensively explained in Chapter 3. The applications of change detection by band algebra and image classification in mine monitoring are presented in two case studies in sections 6.6 and 6.7.

6.6　CASE STUDY I – MINE ENVIRONMENT MONITORING BY CHANGE DETECTION USING BAND ALGEBRA

6.6.1　Study area

This case study was conducted by Nuray Demirel during her M.Sc. studies. All the data, figures and tables presented in this section were taken from her dissertation. The study area, Bolu-Göynük Himmetoğlu Open Cast Mine (GÖLİ), is one of the open cast lignite coal mines belonging to Turkish Coal Enterprises (TCE). GÖLİ was established in 1979 as a branch of Bolu Lignite Administration. The headquarters of the management is in Göynük, which is 30 km away from the city of Bolu and covers an area of about 48 km^2 (4800 hectares). The thickness of the coal bed increases in the easterly direction and the proven lignite coal reserve is 39 million tons. The average annual production is around 170,000 tons of coal. This area is covered by 1:25,000 scaled topographical maps having numbers of 54 H 25 b4 and 54 H 25 c1. The geographical location of the study area in terms of its location in Turkey and Bolu is given in Figure 6.9. It is surrounded by high hills from south to north namely: Aladağ Hill (813 m), Deliklikaya Hill (785 m), and Orman Hill (809 m) to the north; Kozalıkbeleni Hill (659 m) and Kocabelen Hill (608 m) to the south; and Değirmen Hill (665 m) to the east.

The nearest village to the mine is Himmetoğlu which is located very close to the mine site. The other villages are Çayköy, Bölücekova, Ahmetbeyler, and Kuyupunarköy. The roads that connect Himmetoğlu to the surrounding counties, Göynük and Ankara are paved and in general open to transportation in all seasons.

The drainage pattern of the study area is formed parallel to the morphology of the area as a dendritic drainage pattern. Generally streams form V-shaped flood clefts from the top of the hills to the basin. The main streams in the study area are Ovaçay Stream, Boyalıçay Stream, Bölücekova Stream, Mehmetağa Stream, and Eskiköy Stream. The streams and their locations with respect to the villages can be seen in Figure 6.10.

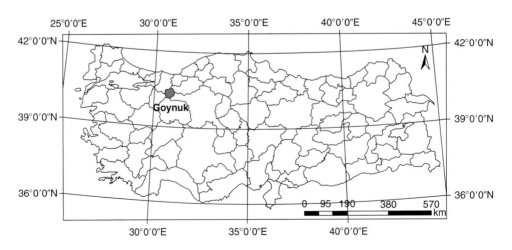

Figure 6.9 Location map of the study area (Erdoğan, 2002)

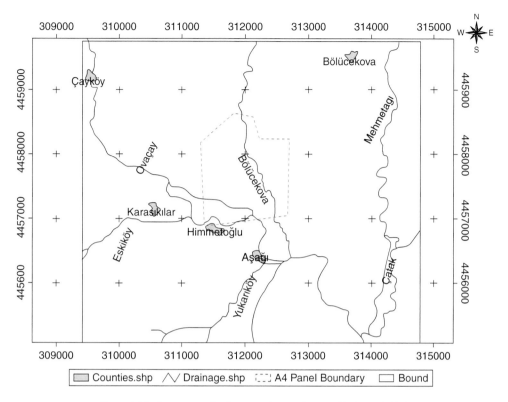

Figure 6.10 Streams and villages in the study area (Erdoğan, 2002)

In the study area, the mining method is an excavator-truck system. The lignite bed is in the form of a narrow corridor lying in the east-west direction, and sharply ascending at the edge of the basin. The mining operation has started from the south of Çayköy and progressed in the easterly direction. Mined out areas were filled with stripped waste material (Figure 6.11) in Kayaboğazı locality. It is proposed that the mine should be expanded in the direction of Bölücekova Stream.

Slopes are generally dipping in an east-west direction parallel to the direction of advance in the south and the north, and in a north-south direction in the direction of advance of mining operations. Excavation equipment consist of 3 hydraulic excavators having bucket capacities of 1.3, 3.0 and 3.2 m^3, 5 backhoes and 1 excavator having a bucket capacity of 2.5 m^3 which is used for excavating the coal horizon after stripping of overburden.

Since there is not enough room for internal dumping, excavated material is dumped outside of the mine site, 22 km away from the mining area, in Kozalıkbeleni Hill to the south of Himmetoğlu and the main dumping area which is near the mine, using 40 ton capacity dumping trucks. Coal is transported to the steelyard of the administration in order to be transferred to the Çayırhan Thermal Power Plant and to the market by means of trucks with capacities of 6 tons. The height of the benches that form the slopes of the mine varies between 5 to 8 m depending on the capacity of excavators

Figure 6.11 Bolu-Göynük open pit lignite mine (Erdoğan, 2002)

being used. The width of the benches varies between 10 to 20 m. Bench faces slopes range from 60° to 65°.

Since the alluvium which forms the top of the overburden material is loose and soft and both the lignite coal bed and the units that belong to the Himmetoğlu formation are weak and fractured by discontinuities, the excavation process is performed easily.

6.6.2 Data

For monitoring of the mine environment in the study area, remote sensing analysis was carried out using SPOT (Système Probatorie d'Observation de la Terre) satellite images. The SPOT system is a system for observation of the Earth and distribution of images acquired by SPOT satellites, which are linked down to ground receiving stations. Acquired images undergo a certain number of routine or optional preprocessing operations before they are utilized by users. SPOT satellites can acquire images in either vertical or oblique viewing.

For the remote sensing analyses three different satellite images were used. They are SPOT Pan for the dates of 08.09.1987 and 02.10.1999 and SPOT XS for the date of 23.09.1999 (Figures 6.12, 6.13, and 6.14). Satellite images have been pre-processed in order to make atmospheric corrections and remove haze effects when they were obtained.

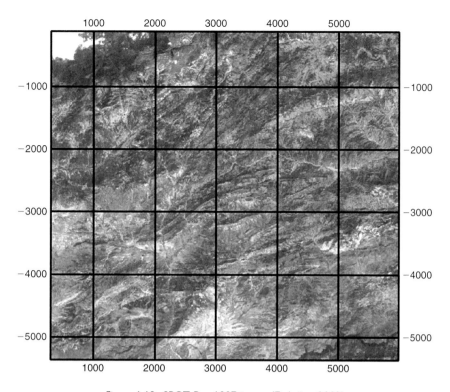

Figure 6.12 SPOT Pan 1987 image (Erdoğan, 2002)

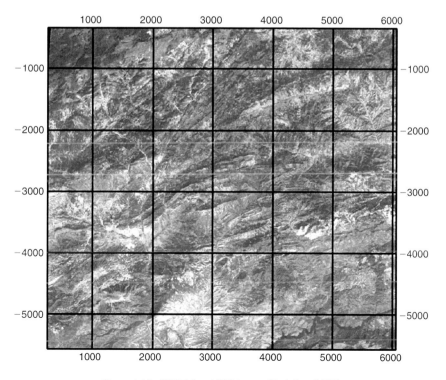

Figure 6.13 SPOT Pan 1999 image (Erdoğan, 2002)

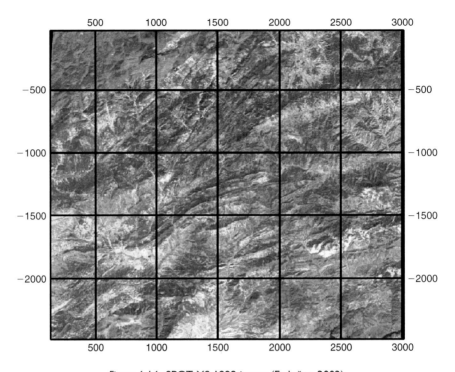

Figure 6.14 SPOT XS 1999 image (Erdoğan, 2002)

As an ancillary data source, 1:25,000 scale topographic maps obtained from the General Command of Mapping were used. Also, 1:5,000 scale contour maps obtained from the Turkish Coal Enterprises were used. Two sheets of 1:25,000 scale topographic maps and 1:5,000 scaled contour maps have been digitized (Figure 6.15). Then the Digital Elevation Model (DEM) from the contour maps were obtained (Figure 6.16), as well as producing slope and aspect maps of the mine site (Figure 6.17 and 6.18).

1:5,000 scale geological maps of the study area were obtained from the General Directorate of Mineral Research and Exploration. These maps were also digitized. Geological characteristics of the study area were given in Figure 6.19. While the elevation contours were being digitized, hill marks, rivers, roads, and residential areas were also digitized as separate layers. According to Figure 6.19, the study area mostly covers geological formations of alluvium, sandstone, marl, and trona.

SPIN 2 satellite image was also used as an ancillary data source in the geometric correction part of the study (Figure 6.20). SPIN-2 satellite image is the trademark for Russian 2-meter resolution orthorectified, panchromatic digital data. Due to its optical design features and the large swath capabilities of the camera systems, highly accurate and detailed imagery can be created for specific regions, or for entire countries as well (Lavrov, 2000). Since the area was partly covered by the frame of the SPIN-2 satellite image, it was not possible to use images of this satellite. However, these data were used as an ancillary data resource for geometric correction purposes. GPS measurements in the field were used in the accuracy assessment part of the classification section.

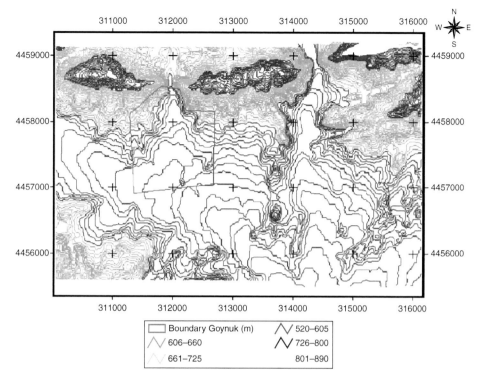

Figure 6.15 Digitized topographic contours of study area (1/25,000 scale) (Erdoğan, 2002) (See color plate section)

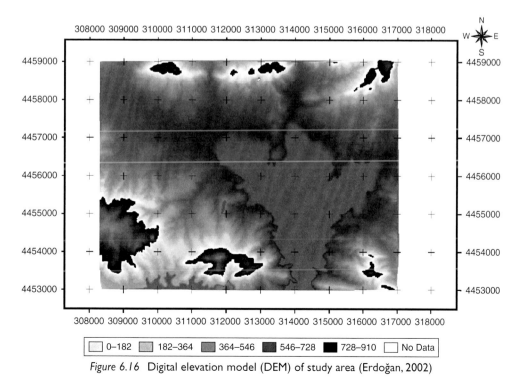

Figure 6.16 Digital elevation model (DEM) of study area (Erdoğan, 2002)

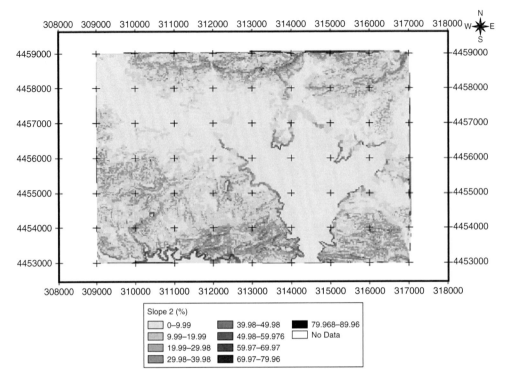

Figure 6.17 Slope map of the study area (Erdoğan, 2002)

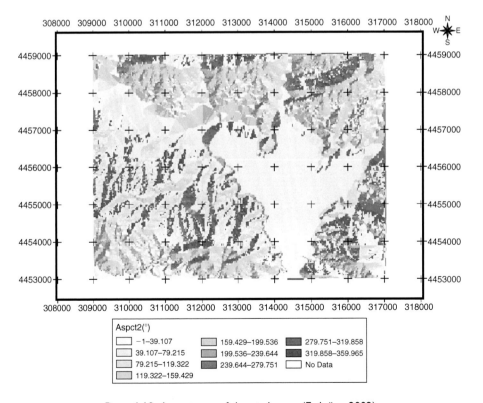

Figure 6.18 Aspect map of the study area (Erdoğan, 2002)

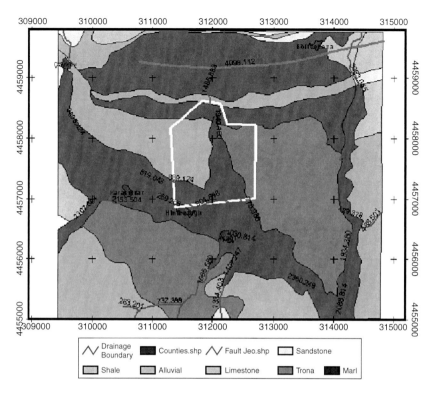

Figure 6.19 Geological map of the study area (Erdoğan, 2002) (See color plate section)

Figure 6.20 SPIN-2 Satellite image as ancillary data (Erdoğan, 2002)

6.6.3   Mine environment monitoring methodology using change detection by band algebra

In this study, change in the topography and land cover was monitored using satellite images. This brings about less labor intensive, less time consuming, more beneficial, and more effective mine monitoring in the long term. In the scope of the study, the areal change of overburden excavation stripping was monitored. For this purpose two medium resolution orthorectified satellite images were used. The approach may give results as near accurate as that of classical methods but it can be developed by additional ground data and topographic measurements. The general steps followed during the pre-processing and processing algorithms of the images are given in Figure 6.21. After acquiring the satellite images and GPS measurements, the images were radiometrically and geometrically corrected. After image preprocessing and image enhancement, in order to determine the change quantitatively, change detection was conducted using band differencing and unsupervised classification algorithms. The obtained results were then utilized to produce change maps of the mine site.

Image preprocessing operations involve radiometric and geometric corrections. Radiometric corrections have been done at Level 1A by the distributor. However, another radiometric correction was performed in order to justify the results of the distributor. Radiometric correction at Level 1A can be defined as application of a linear model to compensate the system error in the data. Level-1A data are intended primarily for users requiring image data that have undergone only minimal preprocessing. After removing radiometric distortions, geometric corrections should be done. The methods and applications of radiometric corrections and geometric corrections of preprocessing were given in the following sections.

Image preprocessing

Multiple date images, which are SPOT panchromatic acquired in 1987 and 1999, were used in this study. In order to eliminate the difference between two satellite data sets caused by atmospheric attenuation, it was necessary to normalize these two images with respect to each other. Due to its simplicity, regression analysis was applied for the purpose of normalization of multiple date images by means of statistical software.

This section of the chapter describes the performed change detection analysis by band differencing algorithm. The change in the region which occurred between the years 1987 and 1999 due to mining activities was detected as an aerial extent. The primary interest in the change detection study was to identify the change which occurred in land cover due to mining activities.

The orthorectified image acquired in 1987 was chosen as a base image because of its clear atmospheric conditions compared to the image acquired in 1999. In order to perform the change detection by means of the band differencing procedure, it was necessary to take the difference of brightness values of each pixel on each band of the two images after normalizing them radiometrically with respect to each other. Theoretically this differencing can be expressed as in Eq. 6.1:

$$D_{ijk} = \left(\frac{DN_{ijk}(t_1) - DN_{ijk}(t_2)}{2} \right) + 127 \tag{6.1}$$

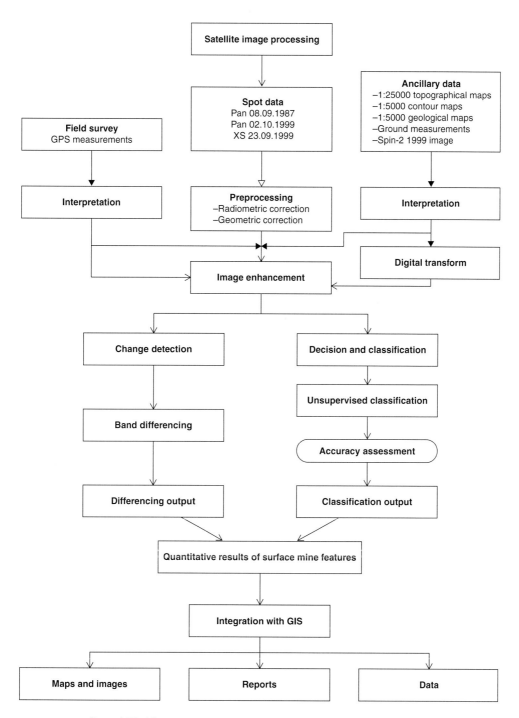

Figure 6.21 Mine environment monitoring methodology (Erdoğan, 2002)

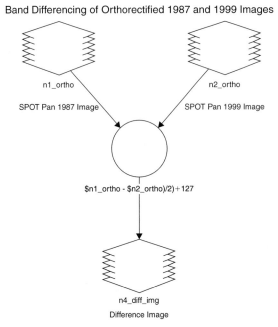

Band Differencing of Orthorectified 1987 and 1999 Images

Figure 6.22 Model of band differencing algorithm (Erdoğan, 2002)

In Eq. 6.1, D_{ijk} is change pixel value, $DN_{ijk}(t_1)$ is brightness value at time 1987, $DN_{ijk}(t_2)$ is brightness value at time 1999, i is line number, j is column number, and k is a single band (e.g. SPOT Pan).

For the purpose of taking the difference of images, the model maker module of the ERDAS Imagine software was used. The obtained model for band differencing is presented in Figure 6.22.

After running the model, the resultant difference image was obtained as in Figure 6.23 for the mine area and Figure 6.24 for the active mine site Panel A4.

Determining optimum threshold values

Most of the methods require a decision as to where to place threshold boundaries in order to separate areas of change from those of no change (Singh, 1989). If an image $I(x, y)$ contains light objects (change) on a dark background (no change), then objects may be extracted by a simple thresholding.

$$I(x, y) = \begin{cases} 1 \Rightarrow I(x, y) > T \\ 0 \Rightarrow I(x, y) \leq T \end{cases} \tag{6.2}$$

In Eq. 6.2, T is the threshold value supplied empirically or statistically by the analyst. All the pixels that belong to the object (change) are coded 1 and the background (no-change) is coded 0. If one wants to define more than one threshold one may use the

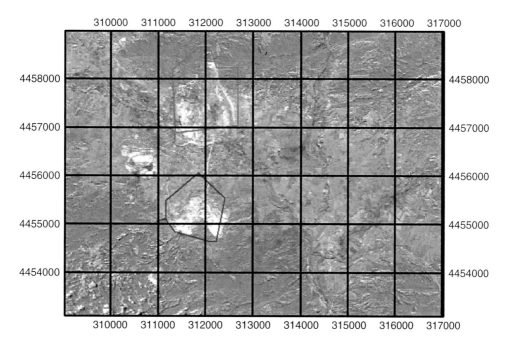

Figure 6.23 The difference image (Erdoğan, 2002)

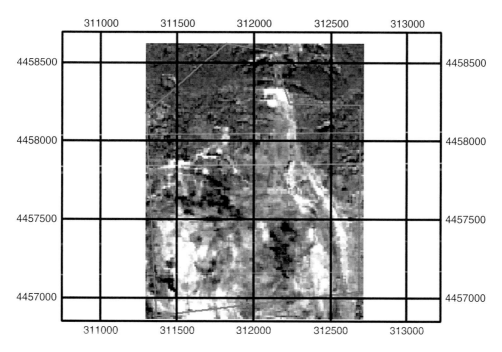

Figure 6.24 The difference image of A4 panel (Erdoğan, 2002)

Table 6.8 Iterative threshold values for different 'N' numbers (Erdoğan, 2002)

N	Upper Threshold	Lower Threshold	Total Area (hectares)	A4 Panel (hectares)
0.4	133.067	119.579	2268.70	93.40
0.6	136.439	116.207	3099.25	123.36
0.8	139.811	112.835	4011.63	159.37
1.0	143.183	106.463	4352.18	176.89
1.2	146.555	106.091	4827.48	209.01
1.4	149.927	102.719	5193.02	222.00
1.6	153.299	99.347	5499.76	231.71
1.8	156.671	95.975	5732.02	240.38

technique of density slicing. In this, several objects of different pixel values are grouped into pre-defined slices. Gray level thresholding can be done interactively with a monitor and operator-controlled cursor, but selection of the best threshold level should usually be associated with a priori knowledge about the scene or visual interpretation. The threshold values may also be derived from the histogram of the image.

It is obvious that the selection of an optimum threshold should be based on the accuracy of classifying the pixels as change or no-change. Various accuracy indices such as overall accuracy, average accuracy and/or combined accuracy can be adopted in determining the threshold levels for change detection. Most indices are biased and affected by the ratio between the numbers of either the reference or classified samples of the change and no-change categories. In general, the use of Kappa coefficient is recommended as the standard measure of accuracy and it takes into account all cells of the error matrices.

In the light of these explanations, the optimum upper threshold (U.T.) and lower threshold (L.T.) values were determined empirically in order to detect the places where changes occur or not according to pixel brightness values, after creating the difference image.

$$U.T. = \mu + (N \times \sigma)$$

$$L.T. = \mu - (N \times \sigma) \tag{6.3}$$

In Eq. 6.3, U.T. is upper threshold value, L.T. is lower threshold value, μ is the mean, σ is the standard deviation, and N is the threshold coefficient.

Fung and LeDrew (1987) stated that the threshold values of N ± standard deviations from the mean can iteratively be selected to separate the change pixels from no-change pixels. In this study, the N value was chosen as 0.4 in the first iteration. In the subsequent iterations, it was increased by 0.2 at each stage until the N value of 1.8 was reached (Table 6.8). When the threshold (N) is increased, the number of change pixels on the image decreases. On the other hand, decreasing N increases the number of change pixels (Table 6.8). Figure 6.25 illustrates the effects of optimum threshold levels on change and no change class boundaries. Visual representation of the effect of N threshold levels on the difference image can be seen in Figure 6.26. For each iteration of N, an accuracy assessment was carried out. For the accuracy assessment,

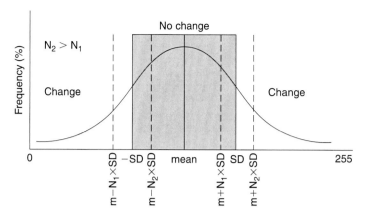

Figure 6.25 The upper and the lower threshold levels defined as boundaries of change and no-change classes (Erdoğan, 2002)

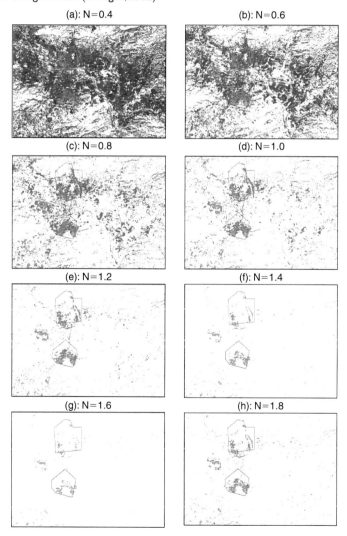

Figure 6.26 Representation of difference images with different threshold values (Erdoğan, 2002)

Table 6.9 Error matrix for N = 0.4 (Erdoğan, 2002)

Difference Image	Reference Image		
	Change	No Change	Row Total
Change	10	103	113
No Change	8	135	143
Column Total	18	238	256

256 samples were collected and the sampling was carried out in a stratified random manner using the accuracy assessment tool of the ERDAS Imagine program.

The accuracy assessment was carried out based on an error matrix, which is the standard form of reporting site-specific error. The random samples were collected on the classified image and each sample was compared with the reference data. According to Hay (1979), at least 50 samples are required for each of the change and no-change classes. In this study, 256 randomly distributed samples were collected.

After distributing the samples, each sample was checked by the corresponding reference data. The boundary of the mine and study area was used as reference data. By comparing each of the samples with the reference data, the change and no-change classes were counted as input values for the error matrix.

The error matrix (Table 6.9) has two rows and columns. The left-hand side (y axis) is labeled with the categories (change-C and no-change-NC) on analyzed classification and the upper edge (x axis) is labeled with the same categories on the reference. The values in the matrix represent the numbers of the pixels (Table 6.9).

An error matrix was derived for each threshold level. After deriving the error matrices, the following four accuracy measurements were computed and the optimum threshold was chosen by comparing them: (i) overall accuracy, (ii) overall kappa, (iii) user's accuracy, and (iv) producer's accuracy.

The results of accuracy criteria and errors are given in Table 6.10. The highest value for each criterion was marked to select the optimum threshold level. Table 6.10 illustrates that the overall accuracy of 90.63 is the highest for the threshold coefficient of 1.2. Besides, when the overall kappa values are considered, 63.13% is the highest that belongs to the same threshold coefficient of 1.2. Furthermore, the average values of both the producer's and user's accuracies for N = 1.2 are almost the highest values among the others. It is important to note here that ERDAS Imagine uses a 95% confidence interval as the default value when computing the user's and producer's accuracies. The graph of kappa index values versus threshold levels was given in Figure 6.27.

For the difference image of orthorectified SPOT images, the optimum 'N' value was found to be 1.2 that gives the most accurate results, and the upper and lower threshold values were found as 147 and 106 respectively. The mean value is 126.323 and standard deviation is 16.860, which means that the pixels that have pixel values between these two threshold values were determined as no-change pixels, and the others had undergone some changes due to mining activities. By doing so, the changes due to mining activities between the years of 1987 and 1999 could be determined easily. The total area that has changes during the 12 years was found to be 48.27 km^2 (4827.48 hectares) for the total mine area and 2.09 km^2 (209.01 hectares) for the A4 Panel.

Table 6.10 Accuracy indices computed from thresholding differenced SPOT Pan image (Erdoğan, 2002)

N Values	Kappa Coefficient (%)	Overall Accuracy (%)	Producer's Accuracy (%)		Omission Errors (%)		User's Accuracy (%)		Commission Errors (%)	
			Change	No Change	Change	No Change	Change	No Change	Change	No Change
0.4	53.43	56.64	55.56	56.72	44.44	43.28	8.85	94.14	91.15	5.59
0.6	24.74	64.06	88.89	60.00	11.11	40.00	26.67	97.06	73.33	2.94
0.8	53.17	81.25	66.67	82.65	33.33	17.35	57.50	92.05	42.50	7.95
1.0	50.17	83.98	60.78	89.76	39.22	10.24	59.62	90.20	40.38	9.80
1.2	63.13	90.63	57.78	97.63	42.22	2.37	83.87	91.56	16.13	8.44
1.4	52.72	89.84	46.15	97.69	53.58	2.30	78.26	90.90	21.74	9.01
1.6	32.33	83.20	26.42	98.03	73.58	1.97	77.78	83.61	22.22	16.39
1.8	40.85	90.23	28.57	100.00	71.43	0.00	100.00	89.84	0.00	10.16

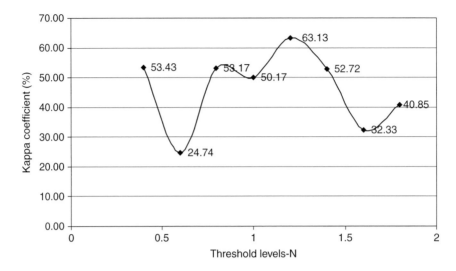

Figure 6.27 Graph of kappa index values versus threshold levels (Erdoğan, 2002)

Therefore, these results proved that the ratio of the changes due to mining activities to the total area was almost 95%. It means that almost entire changes in the region during the 12 year period was caused by mining activities, the rest of the changes may be due to the rural developments in the region, for example, new constructions in rural areas and other environmental variances for example the change in drainage pattern, landslides, artificial lakes, etc. Most of the changes in the A4 Panel occurred in the eastern part of the panel starting from the eastern boundary, which means that in the long-term mining progress would possibly be in the westerly direction. The information provided by TCE verified the obtained result about the progress of mining in the long term in the region.

6.7 CASE STUDY II – MINE ENVIRONMENT MONITORING BY CHANGE DETECTION USING IMAGE CLASSIFICATION

This case study was conducted by Mustafa Kemal Emil under the supervision of H. Şebnem Düzgün in the Mining Engineering Department of Middle East Technical University in Turkey in 2010. All the data, figures and tables presented in this section were taken from Mustafa Kemal Emil's dissertation with the permission of him and his supervisor. The detailed study area description is given in the case study part of Chapter 5.

6.7.1 Data

Data processing steps refers to all pre-processing steps to generate a dataset used in land use land cover analysis. The main data types are pre-processed ASTER

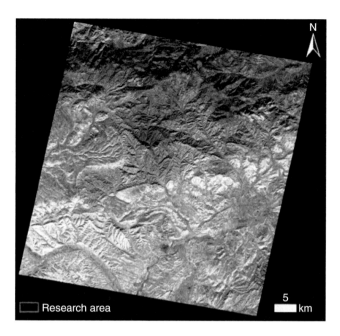

Figure 6.28 Full scene ASTER imagery (False color composite, RGB: 3N21) acquired on November 17, 2002 (Emil, 2010) (See color plate section)

(Figure 6.28), Landsat imagery for land use land cover change detection and Worldview-1 for accuracy assessment due to its high spatial resolution.

6.7.2 Methodology of Land Use and Land Cover (LULC) mapping for mine environment monitoring

Land cover is defined by Comber *et al.* (2005) as the physical material at the surface of the Earth. Land cover consists of forest, bare lands, grass, and water bodies. On the other hand, land use is defined differently by natural and social scientists. According to natural scientists, land use is the modification of natural environment by humanity into the built environment such as agricultural lands and settlements. Social scientists broaden the definition of land use to include the social and economic purposes and contexts of management or – left unmanaged – of land subsistence versus commercial agriculture, rented vs. owned, or private vs. public land (Ellis, 2007). It is important to map Land Use and Land Cover (LULC) for the past and the current stages of the mine site and to observe the changes induced by mining activity. In this study, supervised image classification was implemented in LULC mapping. In the supervised classification method the ASTER dataset was used. The aim of this method was to investigate the efficiency of the ASTER dataset to classify LULC of an abandoned mine site in 2002 and in 2007.

After preprocessing of the data, post-classification comparison was performed to determine changes between pre- and post-mining stages of the site. Mining induced

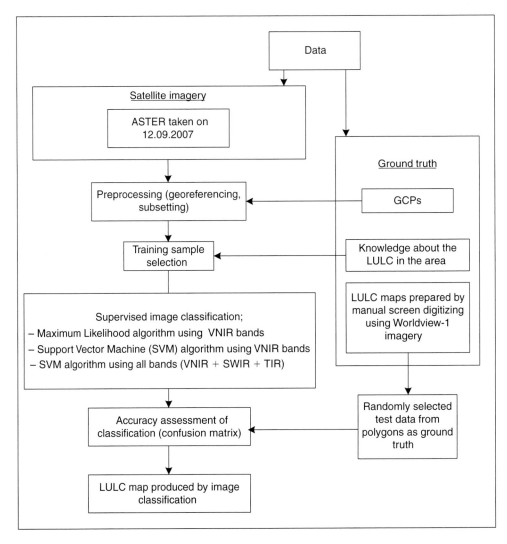

Figure 6.29 Supervised image classification procedure for ASTER dataset (Emil, 2010)

changes in LULC were detected using the intersect tool in the ArcGIS software environment using LULC maps produced for pre- and post-mining stages of the area.

Supervised image classification of ASTER imagery taken in 2007 was performed in the ENVI 4.7 software environment. The ASTER dataset was obtained as orthorectified in UTM projection with WGS1984 datum. During supervised image classification of ASTER imagery, the procedure given in Figure 6.29 was followed.

In this procedure, the first step was the preprocessing of the imagery set. In this step, the ASTER 2007 dataset was georeferenced with collected GCPs. The nearest neighbor resampling algorithm was preferred in the georeferencing process. Moreover, subsets were extracted.

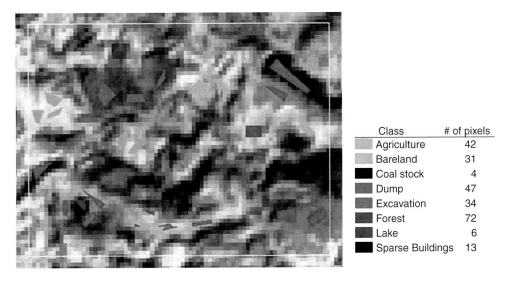

Class	# of pixels
Agriculture	42
Bareland	31
Coal stock	4
Dump	47
Excavation	34
Forest	72
Lake	6
Sparse Buildings	13

Figure 6.30 Distribution and number of pixels of training sample for classification of ASTER imagery taken in 2007 (The background image – false color RGB: 321 – was enhanced by histogram equalization) (Emil, 2010) (See color plate section)

The second step was the selection of training samples which were representative and typical for each of the LULC classes. The selection was based on the field observations, interpretation of various color composites of ASTER imagery, and the Worldview-1 imagery as an ancillary data source. Figure 6.30 shows the distribution and number of pixels for each class of the training sample used in supervised classification of ASTER imagery taken in 2007.

The third step was the implementation of two different supervised classification algorithms, Maximum Likelihood (ML) and Support Vector Machine (SVM). In this step, three different classification procedures were performed: ML using VNIR bands, SVM using VNIR bands, and SVM using all bands (VNIR + SWIR + TIR) for ASTER imagery taken in 2007. The aim of combining bands from different sensors was to integrate the data in order to obtain more information that cannot be obtained from single sensor data alone.

In ML classification the probability threshold was set to zero so that all the pixels were forced to assign to one of the classes. During implementation of SVM classification, the default parameters of ENVI 4.7 were used as the kernel of the radial basis function. Three different classified images were obtained at the end of the image classification step illustrated in Figure 6.31. As seen in Figure 6.31, the result of ML classification has a noisier appearance than the results of the SVM classification. It was observed that increasing the number of bands in the SVM classifier decreased the noisy appearance in the classification results.

The fourth step was the accuracy assessment of the classification results. Accuracy assessment is a process of assessing classification results by comparing ground truth. The most common method of showing classification accuracy is to prepare a confusion

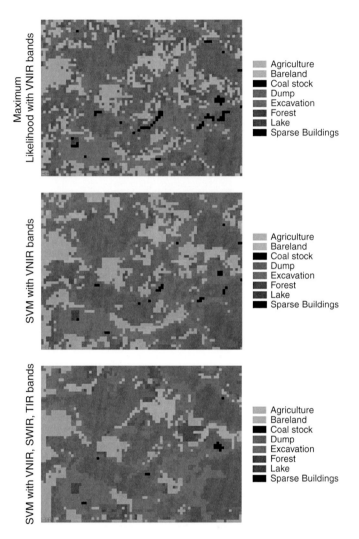

Figure 6.31 Classification results of ASTER dataset by Maximum Likelihood and Support Vector Machine (SVM) classifiers (Emil, 2010) (See color plate section)

matrix (error matrix). Once the confusion matrix was produced, Overall Accuracy, Producer's Accuracy and User's Accuracy for each class could be computed. Another method of assessing accuracy of the classification is to calculate Kappa statistics, which is a measure of the actual agreement between the reference data and an automated classifier and the chance of agreement between the reference data and a random classifier (Lillesand, 2004). In order to assess the classification results, a test sample was required as ground truth. The required test sample was prepared by randomly selecting 30% of each class from the LULC map produced by manual screen digitizing of Worlview-1 imagery. The distribution and pixel numbers of the test sample for each class were

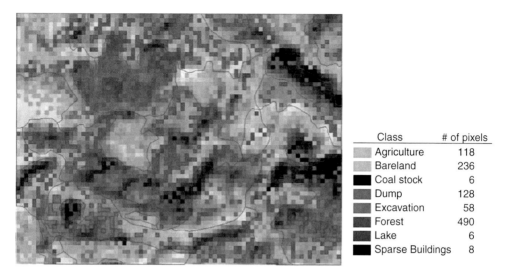

Class	# of pixels
Agriculture	118
Bareland	236
Coal stock	6
Dump	128
Excavation	58
Forest	490
Lake	6
Sparse Buildings	8

Figure 6.32 Distribution and number of randomly selected test pixels (Emil, 2010) (See color plate section)

given in Figure 6.32. The confusion matrices, Overall Accuracy, Producer's Accuracy, User's Accuracy, and Kappa statistics of three classification results were displayed in Figure 6.33.

As seen in Figure 6.33, the overall accuracy of ML classification for VNIR bands of the ASTER dataset taken in 2007 was 41%. The Forest and Agriculture classes were extracted without misclassification with relatively higher Producer's and User's accuracies. The other classes such as Sparse Buildings, Lakes and Coal were prone to some amount of misclassification with relatively lower Producer's and User's accuracies. This was perhaps due to the smaller areal extent of these classes which caused an insufficient set of training sample and mixed spectral reflectance at the boundaries. In the case of SVM classification with VNIR bands, the overall accuracy was 33%. It was observed that distinguishing Bareland and Excavation classes was poor showing lower Producer's and User's Accuracies due to similar spectral signatures in both ML and SVM classifications. Adding the SWIR and TIR bands to the SVM classification procedure apparently increased the classification accuracy. Overall accuracy increased to 39%, showing enhancement in both Producer's and User's Accuracy for all classes. The overall classification accuracies were generally low due to low accuracy of some individual classes such as lakes and sparse buildings which have a very small areal extent with respect to pixel size of the satellite imagery. This also reduced the overall accuracy.

The LULC map was extracted for the above-mentioned classes using VNIR bands of the ASTER dataset taken in 2007 with an overall accuracy of 41% using ML classification. Although ML classification gave slightly better accuracy (41%) than SVM classification (39%), visual inspection of results showed that SVM classification created a more homogenous LULC map than the ML classification. Moreover, it was

(a)

Class	D	A	F	C	L	S	B	E	T	U. A.
				Ground Truth (pixels)						
D	67	4	126	0	0	0	44	0	241	28
A	6	71	54	0	0	3	17	1	152	47
F	35	8	183	0	0	0	11	0	237	77
C	2	0	3	1	3	0	2	0	11	9
L	0	0	0	0	2	0	0	0	2	100
S	3	22	58	0	0	3	24	0	110	3
B	5	1	35	0	0	0	55	6	102	54
E	10	12	31	5	1	2	83	51	195	26
T	128	118	490	6	6	8	236	58	1050	
P. A.	52	60	37	17	33	38	23	88		
Overall Accuracy = 41% Kappa Coefficient = 0.2812										

Confusion matrix for ML classification of ASTER dataset taken in 2007

(b)

Class	D	A	F	C	L	S	B	E	T	U. A.
				Ground Truth (pixels)						
D	69	17	208	0	0	1	55	0	350	20
A	7	80	112	0	0	3	38	1	241	33
F	35	2	112	0	0	0	6	0	155	72
C	1	0	1	0	1	0	2	0	5	0
L	0	0	0	0	4	0	0	0	4	100
S	0	9	21	0	0	2	2	0	34	6
B	0	0	3	0	0	2	33	6	44	75
E	16	10	33	6	1	0	100	51	217	24
T	128	118	490	6	6	8	236	58	1050	
P. A.	54	68	23	0	67	25	14	88		
Overall Accuracy = 33% Kappa Coefficient = 0.2108										

SVM classification using VNIR bands of ASTER dataset taken in 2007

(c)

Class	D	A	F	C	L	S	B	E	T	U. A.
				Ground Truth (pixels)						
D	89	19	192	1	0	0	49	2	352	25
A	8	70	42	0	0	1	10	4	135	52
F	17	3	134	0	0	0	14	0	168	80
C	1	0	1	1	0	0	1	0	4	25
L	0	0	0	0	6	0	0	0	6	100
S	0	7	28	0	0	4	7	0	46	9
B	5	1	14	0	0	0	53	0	73	73
E	8	18	79	4	0	3	102	52	266	20
T	128	118	490	6	6	8	236	58	1050	
P. A.	70	59	27	17	100	50	22	90		
Overall Accuracy = 39% Kappa Coefficient = 0.2733										

SVM classification using VNIR + SWIR + TIR bands of ASTER dataset taken in 2007

Agriculture (A); Dump (D); Forest (F); Bareland (B); Excavation (E); Sparse Buildings (S); Coal Stock (C); Lake (L); Total (T); Production Accuracy in % (P.A.); User Accuracy in % (U.A)

Figure 6.33 Accuracy assessments of image classification results (Emil, 2010)

observed that adding SWIR and TIR bands to VNIR bands in SVM classification of the ASTER imagery increased the overall classification accuracy from 33% to 39%.

REFERENCES

Anderson, A.T., Schultz, D., Buchman, N. & Nock, H.M. (1977) Landsat imagery for surface mine inventory. *Photogrammetric Engineering and Remote Sensing*, 43(8), 1027–1036.

Comber, A., Fisher, P. & Wadsworth, R. (2005) What is land cover. *Environment and Planning B: Planning and Design*, 32(2), 199–209.

Dobson, J.E., Bright, E.A., Ferguson, R.L., Field, D.W., Wood, L.L., Haddad, K.D., Iredale, H., Jensen, J.R., Klemas, V.V., Orth, R.J. & Thomas, J.P. (1995) *Coastal Change Analysis Program (C-CAP): Guidance for regional implementation*. NOAA Technical Report NMFS 123, National Marine Fisheries Service, Seattle, Washington.

Ellis, E. & Pontius, R. (2007) Land-use and land-cover change. *Encyclopedia of Earth*. http://www.eoearth.org/article/Causes_of_land-use_and_land-cover_change (Accessed 16 July 2010).

Emil, M.K. (2010) *Land Degradation Assessment for an Abandoned Coal Mine with Geospatial Information Technologies*. M.Sc. Thesis, Middle East Technical University, Ankara, Turkey.

Erdoğan, N. (2002) *Monitoring Changes in Surface Mining Area by Using SPOT Satellite Images*. M.Sc. Thesis, Middle East Technical University, Ankara, Turkey.

Frimpong, S. (2003) *MINE 413 Surface Mining Methods – Theory and Applications for Undergraduate Mining Engineering Curriculum*. University of Alberta, School of Mining and Petroleum Engineering, Unpublished.

Fung, T. & LeDrew, E. (1987) Application of principal component analysis to change detection. *Photogrammetric Engineering and Remote Sensing*, 53(12), 1649–1658.

Hartland-Swann, J.K. (1993) *The Role of the Landscape Profession in Opencast Coal Mining*. Dissertation, University of Manchester, UK.

Hay, A. M. (1979) Sampling designs to test land use map accuracy. *Photogrammetric Engineering and Remote Sensing*, 42, 671–677.

Hossner, L.R. (1988) *Reclamation of Surface Mined-Lands*, Volume 1. Boca Raton, Florida, CRC Press, Inc.

Jensen, J.R., Cowen, D.J., Narumalani, S., Althausen, J.D. & Weatherbee, O. (1993) An evaluation of coast watch change detection protocol in South Carolina. *Photogrammetric Engineering and Remote Sensing*, 59(6), 1039–1046.

Kaufmann, R.K. & Seto, K.C. (2001) Change detection, accuracy and bias in a sequential analysis of Landsat imagery in the Pearl River Delta, China: Econometric techniques. *Agriculture, Ecosystems and Environment*, 85, 95–105.

Lavrov, V.N. (2000) *Spin-2 Space Survey Photocameras for Cartographic Purposes*. Sovinform-sputnik.

Lillesand, T. M., Kiefer, R. W. & Chipman, J. W. (2004) *Remote Sensing and Image Interpretation* (5 ed.). New York: Wiley.

Lu, D. & Weng, Q. (2007) A survey of image classification methods and techniques for improving classification performance. *International Journal of Remote Sensing*, 28(5), 823–870.

Lu, D., Mausel, P., Brondi, Z.E. & Moran, E. (2005) Change detection techniques. *International Journal of Remote Sensing*, 25(12), 2365–2407.

Muir, B.G. (1979) Observations of wind-blown super-phosphate in native vegetation. *Western Australian Naturalists*, 14, 128–130.

Munshower, F.F. (1983) *Problems in Reclamation Planning and Design, in Coal Development: Collected Papers*, V. II, Bureau of Land Management, U.S. Department of Interior, Denver,

1287. Proc. Symp. Surface Coal Mining and Reclamation in the Northern Great Plains, Mont. Agric. Exp. Stn. Res. Rep. No. 194, Montana State University, Bozeman.

Nicholson, D.T. (1988) *The Role of Landscape Design in Limestone Quarrying with Special Reference to the Mitigation of Environmental Impacts*. Dissertation, University of Manchester, UK.

Singh, A. (1989) "*Digital Change Detection Techniques Using Remotely-Sensed Data*", *International Journal of Remote Sensing*, 6(6), 883–896.

Thunnissen, H.A.M. & Nieuwenhuis, G.J.A. (1989) An application of remote sensing and soil water balance simulation models to determine the effect of groundwater extraction on crop evapotranspiration. *Agricultural Water Management*, 15(4), 315–332.

UNEP (1983) *United Nations Environment Programme*. Report to the Governing Council, General Assembly, Official Records, 38 Session, Supplement No: 25, A/38/25.

Wood, G.A., Loveland, P.J. & Kibblewhite, M.G. (2004) *The Use of Remote Sensing to Deliver Soil Monitoring a Report Prepared for the Department of the Environment, Food and Rural Affairs*, http://www.defra.gov.uk/environment/quality/land/soil/research/indicators/documents/soil-remotesensing.pdf (Accessed 2 January 2011).

Zhengfu, B. & Dazhi, G. (1990) Reclamation, treatment and utilization of coal mining wastes (Rainbow, A.K.M. ed.), *Proceedings of the Third International Symposium on the Reclamation, Treatment and Utilization of Coal Mining Wastes*, Glasgow, United Kingdom, 3–7 September 1990.

Appendix

Table A.1 Technical specifications of Earth observing satellites

Sensor/Satellite	Operator/Country	Launch and End Date	Sensor Type	Resolution		Radiometric (bit)	Swath (km)	Revisit (days)
				Spectral Resolution (μm) or Polarimetry	Spatial (m)			
ALI/EO-1	NASA/USA	2000	PAN Multi	0.48–0.69 / 0.433–0.453, 0.45–0.515, 0.525–0.605, 0.63–0.69, 0.775–0.805, 0.845–0.89, 1.2–1.3, 1.55–1.75, 2.08–2.35	10 / 30	16 / 16	37 / 37	16 (7–9)
ALOS	JAXA/Japan	2006	PALSAR / PRISM / AVNIR/2	L-Band 1270 MHz, 23.6 cm / Pan: 0.52–0.77 (Triplet) / 0.42–0.50, 0.52–0.60, 0.61–0.69, 0.76–0.89	10–100 / 2.5 / 10	3/5 / 7 / 8	20–350 / 70 (nadir) 35 (back) / 70 (nadir)	46 (2)
ALSAT-1	DMC-Algeria	2002	Multi	0.52–0.62 (Green), 0.63–0.69 (Red), 0.76–0.9 (NIR)	32	10	640	4
ASTER/TERRA	METI & NASA Japan-USA	1999	VNIR / SWIR / TIR	0.52–0.60, 0.63–0.69, 0.76–0.86 (S) / 1.60–1.70, 2.145–2.185, 2.185–2.225, 2.235–2.285, 2.295–2.365, 2.360–2.430 / 8.125–8.475, 8.475–8.825, 8.925–9.275, 10.25–10.95, 10.95–11.65	15 / 30 / 90	8 / 8 / 12	60	48 (16)
BEIJING-1	BLMIT/DMC/China	2005	PAN / Multi	0.5–0.8 / 0.52–0.62, 0.63–0.69, 0.76–0.9	4 / 32	8 / 10	24 / 640	14 / 5
BILSAT	TUBITAK/SPACE DMC/Turkey	2003–2006	PAN / Multi	0.45–0.90 / 0.45–0.52, 0.52–0.60, 0.63–0.69, NIR: 0.76–0.90	12.6 / 27.6	8 / 8	25 / 55	116 / 52
CARTOSTAT-1 (IRS-P5)	ISRO/India	2005	PAN	0.5–0.85 (S)	2.5	10	27.5 (S)/55 (M)	126 (5)
CARTOSTAT-2	ISRO/India	2007	Mono/Paint/Brush Multi-View	0.45–0.85	0.8	10	(9.6/28/50) × 9.6 / 9.6 × 38, 28 × 29, 50 × 19 / (9.6/15/28) × 9.6	4–5

Satellite	Operator/Country	Launch	Sensor	Spectral bands (μm)	Resolution (m)	Radiometric	Swath (km)	Revisit
CBERS-1/2	INPE-CAST Brasil-China	1999/2003	PAN	0.51–0.73 (S)	20	8	113	26 (3)
			Multi	0.45–0.53, 0.52–0.59, 0.63–0.69, NIR: 0.77–0.89	20	8	113	26 (3)
			IRMSS	0.50–1.10 (P), 1.55–1.75, 2.08–2.35 (S), 10.40–12.50 (T)	80/160 (T)	–	120	26
CBERS-2B		2007	WFI-Multi	0.63–0.69, NIR: 0.77–0.89	260	8	890	5
			HRC	0.50–0.80 (Additional for Cybers-2B)	2.7	8	27	130
COSMOSKY-MED-1, 2&3	Agenzia Spaziale Italy	2007/2008/ 2009	SpotLight1	X-Band λ: 0.031 m FRQ: 9.6 GHz Gizli: Classified	–	–	–	16 (1)
			SpotLight2	Single: HH / HV	1	3	10	
			PingPong	Dual: HH+VV/	15	8	30	
			StripMap	HH+HV/VV+VH	3–15	3	40	
			WideRegion		30	8	100	
			HugeRegion		100	8	200	
DEIMOS-1	Deimos/ Spain	2009	MS	0.52–0.60, 0.63–0.69, 0.77–0.90	22	8	600	1
DUBAISAT-1	EIAST/ UAE	2009	PAN	0.42–0.89 (S CT)	2.5	8	20	5 (3)
			MS	0.42–0.51, 0.51–0.58, 0.60–0.72, 0.76–0.89 (S-CT)	5	8	20	5 (3)
ENVISAT	ESA/EU	2002	ASAR-Image	C-Band λ: 0.056 m FRQ: 5.33 GHz Single: HH/VV	30	1.5–3.5	56–100	35 (3)
			ASAR-Alt. Pol.	Alternate: HH+HV/ VV+VH/HH+VV	30		56–100	
			ASAR-WS.	Single: HH/VV	150		405	
			MERIS	15 bands (0.39–1.040)	300/1200	8	1150	
EROS A1	ImageSat/ Israel	2000	PAN	0.5–0.9 (S)	1.9/0.9	10	14/7	1.8–4
EROS B1	ImageSat/ Israel	2006	PAN	0.5–0.9 (S)	0.7	10	7	1.8–4
ERS-1/2	ESA/EU	91–00/95	SAR Image Mode	C-Band λ: Frq: 5.3 GHz Single: VV	30	–	100	3–35
FORMOSAT	NSPO/ Taiwan	2004	PAN	0.45–0.90	2	8	24	1
			Multi	0.45–0.52, 0.52–0.60, 0.63–0.69, 0.76–0.90	8	8	24	1

(Continued)

Table A.1 Continued

Sensor/ Satellite	Operator/ Country	Launch and End Date	Sensor Type	Resolution		Radiometric (bit)	Swath (km)	Revisit (days)
				Spectral Resolution (μm) or Polarimetry	Spatial (m)			
GEOEYE-1	GeoEye/USA	2008	PAN Multi	0.45–0.80 0.45–0.51, 0.51–0.58, 0.655–0.69, 0.78–0.92				
HYPERION/ EO-1	NASA/ Taiwan	2004	Hyperion	220 Spectral band (0.4–2.5 um arasi-hyperspectral)	30	16	7.7	16 (7–9)
IKONOS-2	GeoEye/USA Space Imaging	1999	PAN Multi	0.526–0.929 (S) 0.445–0.516, 0.516–0.595, 0.632–0.698, 0.757–0.853 (S)	0.82 3.2	11 11	11.3 11.3	
IRS 1C/D	ISRO/India	1995/1997	PAN LISS-III WiFS	0.5–0.75 (S) VNIR: 0.52–0.59, 0.62–0.68, 0.77–0.86 SWIR: 1.55–1.70, 0.62–0.68, 0.77–0.86	5.8 23 188	6 7 7	70.5 141 812	24/25
IRS-P6 (ResourceSat-1)	ISRO/India	2003	LISS-III LISS-IV AWiFS	0.52–0.59, 0.62–0.68, 0.77–0.86, SWIR: 1.55–1.70 MX: 0.52–0.59, 0.62–0.68, 0.77–0.86 / Mono: 0.62–0.68 0.52–0.59, 0.62–0.68, 0.77–0.86, SWIR: 1.55–1.70	23.5 5.8 56–70	7–10 SWIR 7 10	140 24/7 (M) 2 × 370	24 5 5
JERS 1	JAXA/Japan	1992–1999	OPS OPS	VNIR: 0.52–0.60, 0.63–0.69, 0.76–0.86 (S) SWIR: 1.60–1.71, 2.01–2.12, 2.13–2.25, 2.27–2.40	18 18	6 6	75 75	44 44
KOPMSAT-1	KARI/S. Korea	1999	EOC OSMI (ocean)	PAN: 0.51–0.73 MS: 0.412–0.443, 0.490–0.555, 0.765–0.865	6.6 1000	8 –	17 800	28 28
KOPMSAT-2	KARI/S. Korea	2006	PAN MSC	0.50–0.90 0.45–0.52, 0.52–0.60, 0.63–0.69, 0.76–0.90	1 4	10 10	15 15	28 (3) 28 (3)
LANDSAT-1/2/3	NASA/EOSAT/ USA	1972/75/78	MSS 1978/ 82/83	VNIR: 0.5–0.6, 0.6–0.7, 0.7–0.8, 0.8–0.11	80	8	180	18
LANDSAT-4/5	NASA/USA	1982/84, 1987/–	TM TM TM	VNIR: 0.45–0.52, 0.52–0.60, 0.63–0.69, 0.76–0.90 SWIR: 1.55–1.75, 2.08–2.35 TIR: 10.42–12.5	30 30 120	8 8 8	183 183 183	16 16 16

Satellite	Agency/Country	Year	Sensor	Spectral bands	Resolution (m)	Bits	Swath (km)	Revisit (days)
LANDSAT-7	NASA/USA	1999–2003 (slc-off)	PAN	0.52–0.9	15	8	185	16
			ETM	VNIR and SWIR: Same as Landsat-5	30	8	185	16
			ETM	TIR: 10.42–12.5 (Low-High gain)	60	8	185	16
MODIS/TERRA	METI-NASA/ JP-USA	1999	MODIS	2 spectral band	250	8	2330	1–2
				7 spectral band	500	8	2330	1–2
				26 spectral band	1000	8	2330	1–2
MONITOR-E	GKNPT & RSA/RU	2005–2007	PAN	0.54–0.84	8	8	96	—
			Multi	0.54–0.59, 0.63–0.67, 0.78–0.90	20	8	160	—
NIGERIASAT-1	Nigeria	2003	Multi	0.52–0.62, 0.63–0.69, 0.76–0.90 (NIR)	32	8	640	—
ORBVIEW-3	GeoEye (ORB IMAGE)/USA	2003–2007	PAN	0.45–0.90	1	11	8	8
			Multi	0.45–0.52, 0.52–0.60, 0.62–0.695, 0.76–0.90 (S)	4	11	8	8
ORBVIEW-2 (SeaStar)	GEOEYE (ORBIMAGE)	1997	SeaWifs	0.402–0.422, 0.433–0.453, 0.480–0.500, 0.500–0.520	4500 GAC[1]	10	1500 GAC	1
			NASA	0.545–0.565, 0.660–0.680, 0.745–0.785, 0.845–0.885	1130 LAC[2]		2800 LAC	
PALSAR/ALOS	JAXA-METI/ Japan	2006	Fine	L-Band λ: 0.236 m FRQ: 1.27 GHz Single: HH/VV	6.25	5	40–70	46 (2)
				Dual: HH + HV/ VV + VH	12.5	5	40–70	
			ScanSAR (1)	Full Polarimetry: HH + HV + VH + VV	25	3/5	20–65	
				Single: HH/VV	50	5	250–350	
			ScanSAR (2)	Single: HH/VV	25	5	250–350	
PLEIADES-1&2	CNES/France	2010/ 2012	PAN	0.480–0.830 (S)	0.5 (0.7)	12	20	26 (1)
			MS	0.43–0.55, 0.49–0.61, 0.60–0.72, 0.75–0.94 (S)	2			
PROBA	ESA	2001	HRC	PAN	8	—	8.2	—
			CHRIS	0.415–1.050 with 5–12 nm spectral res. 19 bands	18	—	14	—
RADARSAT-1	CSA/Canada	1995	SAR	C-Band λ: 0.057 m FRQ: 5.3 GHz HH Pol. 16 beam mode (S)	8	—	50–500	24 (3–35)
					18	—		
RADARSAT-2	CSA-MDA/ Canada	2007	SAR	C-Band λ: 0.055 m FRQ: 5.405 GHz HH,VV,HV,VH (S) Single, Dual ve Quad Pol. (S)	3–100	—	20–500	24 (3–35)

(Continued)

Table A.1 Continued

Sensor/ Satellite	Operator/ Country	Launch and End Date	Sensor Type	Resolution Spectral Resolution (μm) or Polarimetry	Spatial (m)	Radiometric (bit)	Swath (km)	Revisit (days)
RAPIDEYE-1/2/3/4/5	RapidEye/DE.	2008	Multi	0.44–0.51, 0.52–0.59, 0.63–0.685, 0.69–0.73, 0.76–0.85	6.5	12	77	5.5 (1)
RASAT	TUBITAK-SPACE/TR	2010	PAN MS	0.42–0.73 0.42–0.55, 0.55–0.58, 0.58–0.73	7.5 15	8 8	30 30	4
RAZAKSAT-1	ATSB & TPM/ Malasia	2009	PAN MS	0.51–0.73 0.45–0.52, 0.52–0.60, 0.63–0.69, 0.76–0.89	2.5 5	8 (10) 8 (10)	20 20	14
RESOURCE-DK1	Ts-SKB/Rus	2006	PAN Multi	0.58–0.85 0.5–0.6, 0.6–0.7, 0.7–0.8	1 2–3	– –	28 28	– –
RISAT-1	ISRO	2009	SAR	C-Band λ: 0.056 m Single, Dual ve FRQ: 5.35 GHz Quad Pol.	3–50	–	30–240	12
SAC-C	CONAE Argentina	2000	MMRS HRTC HSTC	5 VNIR PAN-VNIR PAN-VNIR	175 35 250	8 8 8	360 90 1000	9 (7) 16 2
SAOCOM-1	CONAE Argentina	2010	SAR	L Band λ: 0.235 m FRQ:1.275 GHz	7–100	–	50–400	16
SPIN 2	RUSSIA 98/05	88/95 KVR-1000/DKI Periodical	MK-4 Pan 0.49–0.59 TK-350	Visible: 0.515–0.565, 0.635–0.690, 0.810–0.900 2–1 Topographic Camera: 0.49–0.59 (S)	15, 8, 15 40 10	Film – 200	150 – 	8
SPOT-1/2/3	CNES & Astrium/ France	1986/90/93 –/–/96	HRV-PAN HRV	0.51–0.73 (S) VNIR: 0.50–0.59, 0.61–0.68, 0.79–0.89 (S)	10 20	8 8	60 60	26 (1–4)
SPOT-4	CNES & Astrium/ France	1998	HRV-PAN HRVIR Veget.	0.61–0.68 (S) VNIR: 0.50–0.59, 0.61–0.68, 0.79–0.89, SWIR: 1.58–1.75 0.43–0.47, 0.61–0.68, 0.78–0.89, 1.58–1.75	10 20 1000	8 8 4/8	60 60 2200	26 (1–4) 26 (1–4) 1

Satellite	Agency/Country	Year	Sensor/Mode	Spectral bands (μm)	Resolution (m)	Quantization (bit)	Swath (km)	Revisit (days)
SPOT-5	CNES & Astrium/France	2002	HRS-PAN	0.49–0.69 (S)	10	8	120	26 (1–4)
			HRG-PAN	0.49–0.69 Super mode (S)	2.5–5	8	60	26 (1–4)
			HRG	VNIR: 0.49–0.61, 0.61–0.68, 0.78–0.89 (S)	10	8	60	26 (1–4)
			HRG	SWIR: 1.58–1.75	20	8	60	26 (1–4)
			Veget.	0.43–0.47, 0.61–0.68, 0.78–0.89, 1.58–1.75	1000	4/8 bit	2250	1
QUICKBIRD	Digital Globe USA	2001	PAN	0.445–0.900 (S)	0.61–0.73	11		
			Multi	VNIR: 0.45–0.52, 0.52–0.60, 0.63–0.69, 0.76–0.89 (S)	2.5–2.9	11		
TANDEM-X	DLR&EADS Astrium/Germany	2010	SAR	Properties are similar to TerraSAR-X in tandem orbit configuration				
TERRASAR-X	DLR&EADS Astrium/Germany	2007	Spotlight	X-Band λ: 0.03 m FRQ: 9.65 GHz Dual (HH+VH), Single (VV/HH)	1	–	5×10	15
			StripMap	Single:VV/HH	3	–	15	
			ScanSAR	Dual: HH+VV/HH + HV/VV + HV Single:VV/HH	5	–	30	
					16	–	100	
THEOS	GISTA/Thailand	2008	PAN	0.45–0.90	2	8	22	26 (14–5)
			MS	0.45–0.52, 0.53–0.60, 0.62–0.69, 0.77–0.90	15	8	90	
TOPSAT	BNSC-UK MoD/UK	2005	PAN	0.50–0.70	2.8	–	17	4
			MS	0.40–0.50, 0.50–0.60, 0.60–0.70	5.6	–	17	(2–2.5)
UK-DMC1	DMCII/UK	2003	MS	0.52–0.62, 0.63–0.69, 0.76–0.90	30–40	8	640	14
UK-DMC2	DMCII/UK	2009	MS	0.52–0.60, 0.63–0.69, 0.77–0.90	22	8	660	–
WORLDVIEW-1	Digital Globe USA	2007	PAN	0.40–0.90 (S)	0.5	11	17.6	5
WORLDVIEW-2	Digital Globe USA	2009	PAN	0.45–0.80 (S)	0.46	11	16.4	7 (1.1)
			MS	0.40–0.45, 0.45–0.51, 0.51–0.58, 0.585–0.625, 0.63–0.69, 0.705–0.745, 0.77–0.895, 0.860–1.04 (S)	1.84	11	16.4	

Subject Index

Color plates